The Less Is More Linear Algebra of Vector Spaces and Matrices

The Less Is More Linear Algebra of Vector Spaces and Matrices

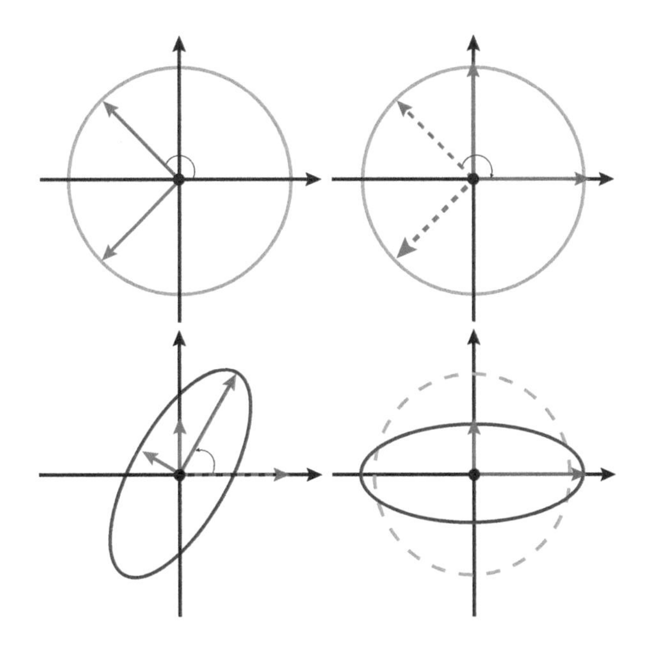

Daniela Calvetti
Case Western Reserve University
Cleveland, Ohio

Erkki Somersalo
Case Western Reserve University
Cleveland, Ohio

Society for Industrial and Applied Mathematics
Philadelphia

10 9 8 7 6 5 4 3 2 1

Publications Director	Kivmars H. Bowling
Executive Editor	Elizabeth Greenspan
Acquisitions Editor	Elizabeth Greenspan
Developmental Editor	Rose Kolassiba
Managing Editor	Kelly Thomas
Production Editor	Ann Manning Allen
Copy Editor	Ann Manning Allen
Production Manager	Donna Witzleben
Production Coordinator	Cally A. Shrader
Compositor	Cheryl Hufnagle
Graphic Designer	Doug Smock

Library of Congress Control Number: 2022948985

In memory of our colleague and friend Elizabeth Meckes

Contents

Preface

Linear algebra is one of the pillars of mathematics and a prototype of it in several ways. It trains the mind to organize ideas and results so as to make the most efficient use of them. Linear algebra is deeply theoretical at its core, yet it plays a crucial role in many extremely practical applications. The power of linear algebra is that it is developed for objects satisfying certain conditions that have nothing to do with computing and numbers, and at the same time it is crucial when it comes to simplifying and streamlining numerical calculations.

In the current era, in which computers have replaced humans when it comes to carrying out complex computational tasks, a solid understanding of linear algebra is more important than ever. Linear algebra is more a way of thinking than a way of computing. This book highlights the concepts of linear algebra and the results that can be obtained by following a rigorous train of thought rather than spending time on tedious calculations that can be handled superbly by computers. While some of the questions addressed will be stated for vectors and matrices with numerical entries, the goal of this book is to train the mind to think abstractly in terms of vector spaces and matrices and to recognize which results are best suited for handling the problem at hand. Adapting these ideas to specific settings where there is a need to address numerical objects will then become a routine application of linear algebra that can often be encoded in an algorithm and solved with the help of a computer. Behind the scenes, linear algebra is the motor propelling the computer tasks. As an outstanding example, the engine that answers our internet searches is powered by linear algebra!

Linear algebra can be very concise, reducing the solution of a complex problem to a very simple one-line answer. Linear algebra is what makes it possible to state whether a linear system has a solution, if the solution is unique or not, and how many qualitatively different solutions can be expected without performing a single calculation.

It was to honor the intrinsic minimalistic beauty of linear algebra and to celebrate the learning ability of our students that made us take on the challenge of writing a rigorous and concise introduction to linear algebra from the point of view of vector spaces and matrices. "Less is more" is what Gene Golub used to say when someone took too long a route to get to the point. We think that those three words describe quite accurately how we chose to present the material in this book: straight to the point with only a few, meaningful illustrative examples. Since linear algebra is not a spectator sport and must be handled to be learned, each chapter ends with a number of problems where readers can test their level of understanding. Additional problems can be found in an accompanying webpage (https://sites.google.com/case.edu/danielacalvetti/less-is-more-linear-algebra) that will be updated over time.

In the spirit of getting in an expedited way to what we believe is the heart of linear algebra, vector spaces are introduced early on, together with inner products and vector norms, to allow us to formally define orthogonality in abstract vector spaces. This is followed by the definition of matrices and matrix norms, which is all else that is needed to set the stage for introducing the singular value decomposition (SVD), arguably the most interesting and useful tool of linear algebra.

The importance of the SVD in applications is hard to overstate, as it has a key role in many areas of applied mathematics, including data science, inverse problems, uncertainty quantification, and imaging. There is no question that anyone working on theoretical or applied mathematical problems cannot afford to be without such a versatile tool as the SVD. The numerous linear algebra based advanced courses in applied mathematics that we have taught over the decades have made us fully appreciate the power and beauty of the SVD in analyzing and understanding linear problems, providing great insight when setting up numerical computations, even when the SVD is not used in the algorithms. This book is our pledge to make the path through linear algebra simpler and smoother for future generations by equipping them with the SVD early on.

The book covers all major linear algebra topics, including Gaussian elimination, solutions of linear systems, determinants, and eigenvalues and eigenvectors. Vector spaces will be present virtually in every chapter throughout the book, providing an interesting and powerful interpretative key for all the results.

The choice of giving vector spaces and the SVD precedence over classical topics like Gaussian elimination and eigenvalues was motivated by the observation that often these two former topics end up being taught at the end of a one semester course, with little time to be absorbed and mastered, and no time to test their theoretical and practical power. We also wanted to make clear that linear algebra is much more than solving linear systems of equations, or computing eigenvalues by finding roots of characteristic polynomials, which many students have already been exposed to in some form or another, and not seldom found rather uninspiring and boringly repetitive. In this book, one has to wait until Chapter 10 before the solution of linear systems via Gaussian elimination is addressed, and at that point vector spaces will be fully operational, providing a bird's eye view of the computations.

The book is intended for a one semester linear algebra course. While it is not assumed that students have already taken calculus, it is expected that students have the maturity needed to learn and appreciate rigorous abstract thinking. The context is organized as follows.

Chapter 1 sets the stage by introducing sets and operations on sets, relations, and functions, and by explaining how sets equipped with certain operations satisfy the axioms of groups and fields, which will be important in the context of vector spaces. The four basic ways to prove a theorem in a rigorous way are also outlined in this chapter. In many vector spaces the scalars are complex numbers, and so are the eigenvalues by default. A comprehensive review of complex numbers with their different representations and their properties is provided in Chapter 2. Chapter 3 is when we get a first real feeling for the heart and soul of linear algebra. In this chapter we generalize the definition of vectors from $\mathbb{R}^2$ and $\mathbb{R}^3$ to more general settings by defining vectors as objects that satisfy a certain list of properties. Vector spaces are the sets of such objects. In the same chapter we also define linear independence and spanning sets, which are needed to define a basis of a vector space. Some fundamental properties of vector spaces and their bases are also presented in Chapter 3.

Inner products among vectors of a vector space and their properties are the topic of Chapter 4. Inner products provide the means to generalize concepts that have a natural concrete geometric interpretation in $\mathbb{R}^2$ and $\mathbb{R}^3$ to abstract vector spaces. Of particular importance and far reaching consequences is the concept of orthogonality among elements of a vector space, defined in terms of an underlying inner product. The two centerpieces of this chapter are the Gram–Schmidt orthogonalization and the Cauchy–Schwarz inequality, which will be used in a variety of contexts throughout the following chapters.

Vector norms, which are the focus of Chapter 5, are ways to assess the size of the vectors in a vector space and can be thought of as generalizations of the length of vectors in $\mathbb{R}^2$ and $\mathbb{R}^3$. Norms can be defined in terms of inner products or by assigning a function that satisfies certain properties. A comparison of three different norms is presented to motivate the choice of one norm over another.

Matrices, first introduced in Chapter 6, have a centerstage role in linear algebra. Matrices are two-dimensional arrays that are susceptible to different interpretations depending on the context in which they are utilized. In many cases matrices provide a convenient way to arrange entries, and hence are viewed "for what they are," while other times when their main role is to operate on vectors, they are considered "for what they do." In this chapter, in addition to defining operations between matrices, we suggest a few different ways to think about matrices, from the natural representations of linear maps between vector spaces to the collections of their rows or columns. The row-wise or columnwise disaggregation of a matrix will be very convenient when we address the solution of linear systems. The definition of the product of a matrix and a vector is an example of how careful one needs to be when working with these two-dimensional arrays because of the need to satisfy compatibility conditions. Matrix-vector products, presented from three different points of view to provide a choice of interpretations to better suit different situations, are extended to define matrix-matrix products, binary operations that combine ordered pairs of compatible matrices, possibly from different vector spaces, to produce a matrix that may belong to neither of the spaces of the parent matrices. After outlining the mechanics of matrix operations, the chapter proceeds to introduce invertibility, orthogonality, matrices with special patterns, and the vector spaces defined by the action of a matrix, most notably range and null space. The chapter ends with the factorization of a matrix as the product of two matrices with special properties, the workhorse of many theoretical arguments and one of the pillars of numerical linear algebra.

Norms of matrices are the subject of Chapter 7. When measuring the size of a matrix, it is possible to choose a norm, the Frobenius norm, that only considers its entries, regardless of their organization into rows and columns, or to select norms that weigh the transformation that the matrix exerts on vectors that are multiplied by it.

Chapter 7 completes all the groundwork needed to introduce the SVD, in our opinion the grandest of all matrix decompositions and certainly the most democratic, because it exists for every matrix, regardless of the number of rows and columns, or other properties of the matrix. Chapter 8 opens with the formal definition of the SVD, which expresses a matrix as the product of three matrices, two orthogonal matrices and a diagonal one with nonnegative entries. The main theorem of this chapter states the universal existence of the SVD and provides an interpretation for the factor matrices. The chapter also presents an alternative formulation of the SVD, where the matrix is expressed as the sum of simple matrices constructed from the columns of the orthogonal factors of the SVD. The SVD is what makes it possible to analyze every matrix as if it were a diagonal matrix. The SVD is used extensively in a wide range of applications and is one of the main engines in many data science algorithms.

Chapter 9 defines the four vector spaces, referred to as the four fundamental subspaces of a given matrix, implicitly associated with every matrix, and derives their properties in terms of the SVD of the matrix. These subspaces play an important role in establishing the solvability of a linear system associated with a matrix. Pairwise orthogonality of the four fundamental subspaces is established, and it is proved that orthonormal bases for these subspaces can be extracted from the matrices of the SVD.

Linear systems of equations and Gaussian elimination are the subjects of Chapter 10. There are a few reasons for postponing this topic, which is many students' first encounter with linear algebra, until this late in the book. One reason for this delayed appearance is to avoid sending the message that linear algebra amounts to the nitty-gritty operations that are part of the elimination process. The second, more compelling reason is that at this point in the book, the higher sophistication level acquired working with vector spaces makes it possible to appreciate the power, elegance, and depth of the procedure and to gain insight into the general principles behind it. Moreover, one of the advantages of analyzing linear systems at this stage is that it is natural to address the topic from the perspective of the SVD and the four fundamental subspaces.

Determinants and their properties are introduced in Chapter 11. The definition of determinant that we present is in terms of multilinear functions, introduced in the beginning of the chapter. Many of the properties of determinants and some of the formulas routinely used for calculating the determinant of 2×2 and 3×3 matrices are derived directly from the definition. The choice of not starting with the combinatorial definition of the determinant that many students are familiar with is again similar to the previous reasoning: linear algebra is a deep theoretical framework that is built not on computations but on elegant and efficient ideas.

Chapter 12 introduces eigenvectors and eigenvalues and proves many important results. Most of the chapter discusses matrices with real entries and establishes the conditions necessary for real matrices to have real eigenvalues and to admit real eigenvectors. A special subsection is dedicated to the characterization of matrices with orthogonal eigenvectors. In this chapter we prove some well-known important results about the spectral factorization.

The book ends with Chapter 13, which is dedicated to solutions of linear systems in the least squares sense. While typically in a linear algebra course square linear systems get most of the attention, in practical applications it is much more likely that one needs to solve linear systems in the least squares sense. Least squares solutions are typically defined for linear systems where the number of equations exceeds the number of unknowns. The chapter starts with the introduction of orthogonal projections, which play an important role in the solutions of least squares problems, and ends with a brief overview of how to actually compute the solution. The details of the numerical solution of least squares problems are not presented, as they are typically topics of a scientific computer course.

The material in this book has been used for several years as the main reference for a one semester proof-based course on linear algebra at Case Western Reserve University. Our students liked its "less is more" style, the focus on concepts and ideas, and problems provided at the end of each chapter and on the accompanying website. They appreciated the directness we used to convey the information and the fact that they did not need to search through an example-laden body of text for the relevant points. The final version has taken into account the feedback of students and colleagues who learned and taught from preliminary versions of this textbook. We owe special thanks to Weihong Guo, Wanda Strychalski, and Daniel Szyld for their valuable comments, suggestions, and corrections. Also, we extend our thanks to the anonymous reviewers of the book for their suggestions on how to improve the presentation of the material. The publication staff at SIAM has been extremely professional and supportive during the various stages that led to the production of this book. We are particularly grateful to Elizabeth Greenspan, whose continuous support and help has been crucial and made working on this project a joy.

Chapter 1

Prelude to Linear Algebra

Linear algebra, as the name suggests, deals with algebraic structures with an emphasis on linear relations between objects. In order to formalize these concepts, we start the exposition with a brief summary of some basic elements constituting the building blocks of the theory.

1.1 ▪ Proof styles

The power of linear algebra comes from the fact that many of its results hold for a wide range of objects, some of which are very concrete and part of our everyday life, and others so abstract so as to elude even the wildest of imaginations. The price to pay for such generality is the need to replace intuition with a tight set of rules to prove what is true. The journey through linear algebra proceeds by defining families of objects that have similar properties and establishing results that hold for all family members. Depending on the content and on the context, these statements are enunciated in the form of lemmas, theorems, or corollaries. Proving a theorem requires showing in a rigorous manner the correctness of its statement. Often theorems are formulated in the form *If P is satisfied, then Q holds.* There are four main types of proofs.

1. **Direct proof**. The process starts from the hypothesis P of the theorems and then advances through a sequence of statements showing that the truth of P implies that Q is true also. Some direct proofs are in the form of procedures and can be implemented in the form of algorithms.

2. **Proof by induction**. Proving that a statement holds *for all n*, where n is a nonnegative integer, is done by first proving directly that the statement holds for the first few values of n, e.g., $n = 1, 2$. Next it is assumed that the statement of the theorem holds for $n = k$, usually referred to as the *induction hypothesis*, and then it is shown that, if this is the case, it also holds for $n = k + 1$. This completes the proof. Some important linear algebra results are proved by induction.

3. **Proof by contraposition**. One of the pillars of logic is that the implication *If P, then Q* is equivalent to *If not Q, then not P*. Therefore if it is not clear how to prove a theorem directly, one may try to prove that the negation of the thesis implies the negation of the hypothesis. This indirect way of proving a theorem is not constructive and may not shed much light on how to take advantage of the theorem in an algorithmic way, but it is formally correct.

4. **Proof by contradiction.** This style of proof starts by negating the thesis and shows that by so doing one arrives at contradicting something that is obviously true, for example proving that $0 = 1$! A proof by contradiction is not constructive but can be an easy way to prove a rather complex statement.

The decision of how to prove a theorem will depend on the theorem to be proved: for some theorems there are very few choices, while others can be proved in more than one way. Sometimes proving a theorem directly may be extremely hard, but it may be very easy to prove it by contradiction.

1.2 ▪ Sets

A set is a collection of objects, referred to as its elements. Sets can be defined by giving a listing all the elements or by stating a property satisfied by the elements. In the words of Georg Cantor, the founder of naive set theory, "a set is a gathering together into a whole of definite, distinct objects of our perception or of our thought, which are called elements of the set." The ensemble of red cars at a dealership is a set, as is the collection of Elvis Presley's jumpsuits. In the mathematical context, the sets that are typically encountered are sets of numbers, and among them some are so common as to be denoted by special symbols. These include the set of natural numbers $\mathbb{N}$,

$$\mathbb{N} = \{0, 1, 2, \ldots\},$$

and the set of integers $\mathbb{Z}$,

$$\mathbb{Z} = \{0, \pm 1, \pm 2, \ldots\}.$$

Some sets contain a finite number of elements, others infinitely many. The sets $\mathbb{N}$ and $\mathbb{Z}$ contain infinitely many elements, and in fact all elements of $\mathbb{N}$ are also elements of $\mathbb{Z}$. Sets that play an important role in linear algebra include the rational numbers,

$$\mathbb{Q} = \left\{ \frac{m}{n} \mid m \in \mathbb{Z}, n \in \mathbb{Z}, n \neq 0 \right\},$$

and the real numbers $\mathbb{R}$.

Given two sets A and B, A is a *subset* of B if every element of A is also an element of B: we denote this relation by

$$A \subset B.$$

For example,

$$\mathbb{N} \subset \mathbb{Z} \subset \mathbb{Q} \subset \mathbb{R}.$$

Two sets A and B are equal if they have the same elements. Thus,

$$A = B \text{ if and only if } A \subset B \text{ and } B \subset A.$$

Observe that the order in which the elements of a set are given is immaterial; the only significant concept is the membership. Therefore, $\{0, 1\} = \{1, 0\}$.

Examples of sets that are defined through a property of the elements and alternatively listing all the elements are

$$A = \{n \in \mathbb{N} \mid 1 \leq n \leq 20\} = \{1, 2, \ldots, 19, 20\}$$

and

$$B = \{n \in \mathbb{N} \mid 5 \leq n \leq 25\} = \{5, 6, \ldots, 24, 25\}.$$

In this case, neither set is contained in the other, but there are some elements that belong to both of them. The *intersection* of A and B is the set whose elements belong to both A and B,

$$A \cap B = \{c \mid c \in A \text{ and } c \in B\}.$$

In the case of the two sets A and B introduced above,

$$A \cap B = \{n \in \mathbb{N} \mid 5 \leq n \leq 20\}.$$

Clearly, the intersection of two sets is a subset of both sets.

The set of all elements that belong to either the set A or the set B, or to both, is the *union* of the sets,

$$A \cup B = \{c \mid c \in A \text{ or } c \in B\}.$$

The union of the two sets A and B defined above is

$$A \cup B = \{c \mid 1 \leq c \leq 25\}.$$

The union of two sets has each set and their intersection as subsets. Observe that we tacitly assume that a set cannot have duplicate elements.

The set theoretic equivalent of the zero as an indication of nothingness is the *empty set*, defined as the set with no elements,

$$\emptyset = \{\ \}.$$

The empty set plays an important role when sets are defined via a property of the elements, because in some cases verifying whether there are any objects with a given trait may be a difficult task. By convention, the empty set is a subset of every set.

The elements belonging to sets are not restricted to numbers, and the operations among sets, such as union or intersection, do not assume anything about the nature of the elements. In linear algebra, the elements of most of the sets of interest are numbers, arrays of numbers, or functions.

We will need the *Cartesian product* of sets: Given two sets A and B, the Cartesian product is defined as a set of ordered pairs,

$$A \times B = \big\{(a, b) \mid a \in A, b \in B\big\}.$$

The Cartesian product can be extended to more than two sets: If $A_1, \ldots, A_n$ are sets, we define

$$A_1 \times A_2 \times \cdots \times A_n = \big\{(a_1, a_2, \ldots, a_n) \mid a_j \in A_j\big\}.$$

1.3 ▪ Functions

Given two sets A and B, a *relation* is defined formally as any set of ordered pairs,

$$\{(a, b) \mid a \in A,\ b \in B\}.$$

Often, a relation is defined through an *assignment*, that is, a rule that associates to the elements of the set A the elements of the target set B. Assignments do not require that the ordered pairs obey any particular condition.

A *function* is an assignment that associates to each element of one set, called the *domain* or *input*, one and only one element of the target set of possible output values, called the *codomain*. Formally, if we denote the input set by A and the target by B, a function is an assignment f,

$$f : A \to B,$$

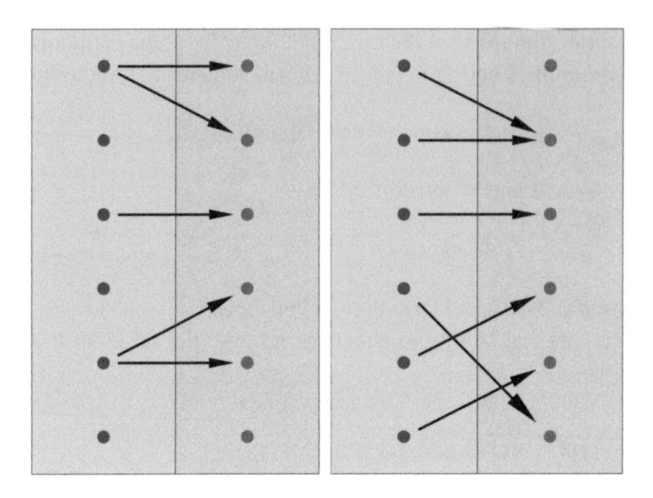

Figure 1.1. *Left panel: A relation that is not a function. Right panel: A relation that is a function. To every element in the domain, exactly one element in the codomain is assigned.*

that associates to each element x of A one and only one element $y = f(x)$ of B,

$$x \mapsto y = f(x).$$

All functions are relations, but not all relations are functions. For example, the assignment

$$f : \mathbb{R} \to \mathbb{R}, \quad f(x) = x^2$$

is a function, but the assignment

$$f : \mathbb{N} \to \mathbb{Z}, \quad f(x) = \pm x$$

is not, because it assigns two possible values to every nonzero element x.

The collection of all elements of B that are assigned to elements of A by the function f is called the *range* of the function. The range is a subset of the codomain.

A function $f : A \to B$ is *injective* if distinct elements of A are assigned to distinct elements of B; i.e., if $x_1 \neq x_2$, then $f(x_1) \neq f(x_2)$. The fact that if $f(x_1) \neq f(x_2)$ then $x_1 \neq x_2$ follows from the definition of function.

A function $f : A \to B$ is *surjective* if for each element $y \in B$ there is an element $x \in A$ such that $f(x) = y$.

A function f is a *bijection* if it is both an injection and a surjection.

Given two functions, $f : A \to B$ and $g : B \to C$ such that the codomain of f is the domain of g, it is possible to define the *composition* of f and g as the function

$$g \circ f : A \to C, \quad g \circ f(x) = g(f(x)).$$

If f is a bijection, the assignment

$$g : B \to A, \qquad g(y) = x \quad \text{if and only if} \quad f(x) = y$$

is also a bijective function. Moreover,

$$g \circ f : A \to A, \quad g \circ f(x) = x, \text{ for all } x \in A,$$

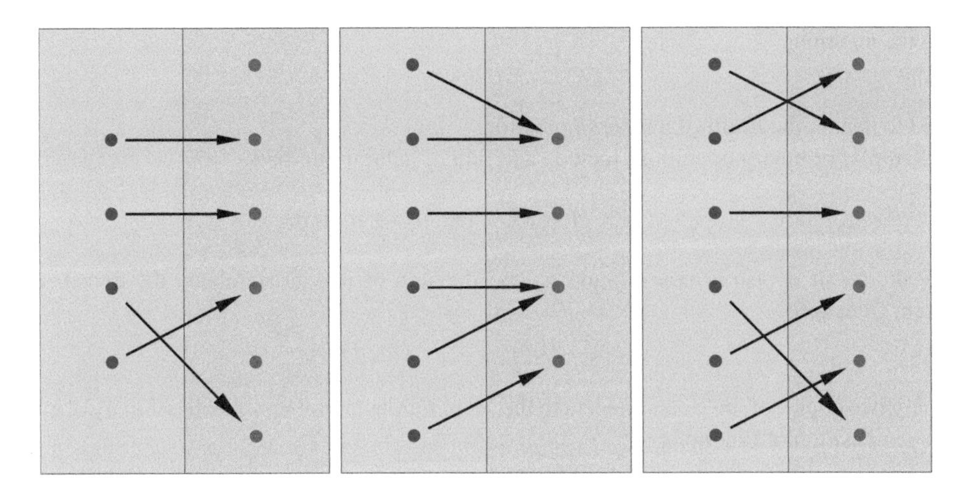

Figure 1.2. *Left panel: An example of injective function. Each element of A is mapped to a different element of B. Middle panel: A surjective function. At least one element of A is mapped to every element of B. Right panel: A bijective function. The elements of A and B are in a one-to-one correspondence.*

and

$$f \circ g : B \to B, \quad f \circ g(y) = y, \text{ for all } y \in B.$$

The function g is called the *inverse* of f, denoted by $g = f^{-1}$, and we have

$$f^{-1} \circ f = \mathrm{id}_A, \quad f \circ f^{-1} = \mathrm{id}_B,$$

where id_A and id_B are the identity functions of A and B, respectively, mapping every element of the set to itself.

1.4 ▪ Operations

A *unary operation* on A is a function from A to B.

Examples of unary operations are the square root of nonnegative real numbers $\mathbb{R}_+ = \{t \in \mathbb{R} \mid t \geq 0\}$,

$$\mathbb{R}_+ \to \mathbb{R}_+, \quad x \mapsto \sqrt{x},$$

and the reciprocal of nonzero integers $\mathbb{Z} \setminus \{0\}$,

$$\mathbb{Z} \setminus \{0\} \to \mathbb{Q}, \quad x \mapsto \frac{1}{x}.$$

A *binary operation*, denoted here symbolically by $\square$, on a pair of sets A_1 and A_2 is a function

$$\square : A_1 \times A_2 \to B$$

that maps an ordered pair of elements, the first from A_1 and the second from A_2, to an element of B,

$$b = \square(a_1, a_2).$$

Often the notation

$$a_1 \square a_2$$

is used to denote the result of a binary operation.

Examples of binary operations include addition of natural numbers,

$$+ : \mathbb{N} \times \mathbb{N} \to \mathbb{N}, \quad (m, n) \mapsto m + n,$$

where the result is also a natural number, and division of integer numbers, the divisor being different from zero,

$$\div : \mathbb{Z} \times (\mathbb{Z} \setminus \{0\}) \to \mathbb{Q}, \quad (m, n) \mapsto \frac{m}{n},$$

mapping two copies of the set of integers to the set of rational numbers. Another binary operation is the composition of functions,

$$\circ : \{f : A \to B\} \times \{g : B \to C\} \to \{h = g \circ f : A \to C\},$$

mapping ordered pairs of functions from A to B and from B to C to the set of functions from A to C. Observe that in the last example, the symbolic order $\circ : (f, g) \mapsto g \circ f$ must be reversed, as f is applied first.

1.5 ▪ Groups, fields, and rings

Let A be a set, and let $\clubsuit$ be a binary operation that associates to an ordered pair of elements of A an element of A.

Definition 1.1. *A set A equipped with a binary operation $\clubsuit$ is a* monoid *if the following axioms are satisfied:*

1. *Closure: For each pair $a, b \in A$ we have $a \clubsuit b \in A$.*
2. *Associativity: For all $a, b, c \in A$, it holds that $a \clubsuit (b \clubsuit c) = (a \clubsuit b) \clubsuit c$.*
3. *Identity: There exists an element $e \in A$ such that for every $a \in A$, $a \clubsuit e = e \clubsuit a = a$.*

The element e above is called the identity, *or* neutral element *of the monoid.*

Definition 1.2. *A set A equipped with the binary operation $\diamond$ is a* group *if the following four axioms are satisfied:*

1. *Closure: For all $a, b \in A$, we have $a \diamond b \in A$.*
2. *Associativity: For all $a, b, c \in A$, it holds that $a \diamond (b \diamond c) = (a \diamond b) \diamond c$.*
3. *Identity: There exists an identity element $e \in A$ such that $a \diamond e = e \diamond a = a$ for all $a \in A$.*
4. *Inverse: For every $a \in A$, there is an element $b \in A$, called the* inverse *of a, such that $a \diamond b = b \diamond a = e$.*

A set A equipped with the binary operation $\diamond$ is an Abelian group *or* commutative group *if it is a group that satisfies the following additional axiom:*

5. *Commutativity: For all $a, b \in A$, it holds that $a \diamond b = b \diamond a$.*

The set of natural numbers $\mathbb{N}$ with the usual addition is a monoid, the number $e = 0$ being the identity, but not a group, because the inverse axiom is not satisfied.

The set of integers $\mathbb{Z}$ with the usual addition is an Abelian group. Again, the identity is $e = 0$, and for any integer a, the inverse $-a$ is contained in the set.

The set of integers $\mathbb{Z}$ with the operation of multiplication is a monoid with the identity element $e = 1$, but not a group, because the multiplicative inverse of each entry is not guaranteed to be in the set.

Definition 1.3. *A set R equipped with two operations $\diamond$ and $\clubsuit$ is a* ring *if the following conditions are satisfied:*

1. *R is an Abelian group with respect to $\diamond$.*
2. *R is a monoid with respect to $\clubsuit$.*
3. *The operation $\clubsuit$ is distributive over $\diamond$ in the following way: For all $a, b, c \in R$, it holds that $a \clubsuit (b \diamond c) = (a \clubsuit b) \diamond (a \clubsuit c)$.*

The set $\mathbb{Z}$ of integers with the usual operations of addition ($\diamond = +$) and multiplication ($\clubsuit = \times$) is a ring.

Definition 1.4. *A set $\mathbb{F}$ equipped with two operations $\heartsuit$ and $\spadesuit$ is a* field *if it satisfies the following conditions:*

1. *$\mathbb{F}$ is an Abelian group with respect to $\heartsuit$.*
2. *$\mathbb{F}$ is a commutative monoid with respect to $\spadesuit$.*
3. *Every element of $\mathbb{F}$ different from the identity of $\heartsuit$ has an inverse with respect to $\spadesuit$.*
4. *The operation $\spadesuit$ is distributive over $\heartsuit$ in the following sense: For every $a, b, c \in \mathbb{F}$, $a \spadesuit (b \heartsuit c) = (a \spadesuit b) \heartsuit (a \spadesuit c)$.*

The set of rational numbers $\mathbb{Q}$ and the set of real numbers $\mathbb{R}$ with the usual addition ($\heartsuit = +$) and multiplication ($\spadesuit = \times$) are fields. For $\mathbb{Q}$ and $\mathbb{R}$, 0 is the additive identity and 1 is the multiplicative identity. It is customary to denote the additive identity by 0 and the additive inverse of x by $-x$, the multiplicative identity by 1, and the multiplicative inverse of $x \neq 0$ by x^{-1} or $\frac{1}{x}$. When convenient and unambiguous, we will use the notation

$$ab^{-1} = \frac{a}{b}.$$

The following theorem shows that in a field the identity elements and the inverses are unambiguous. We will use the notation $+$ and $\times$ for the additive and multiplicative operations in the field.

Theorem 1.5. *In a field $\mathbb{F}$,*

1. *The additive identity, denoted by 0, is unique.*
2. *The multiplicative identity, denoted by 1, is unique, and is different from the additive identity.*
3. *For any $x \in \mathbb{F}$, the additive inverse $-x$ is unique.*

4. *For any $x \neq 0$, the multiplicative inverse x^{-1} is unique, and is different from the additive identity.*

5. *For all $x \in \mathbb{F}$, $0 \times x = 0$.*

6. *For all $x \in \mathbb{F}$, $(-1) \times x = -x$.*

Proof. If 0 and $\widetilde{0}$ are two additive identities, then for all $x, y \in \mathbb{F}$, we have

$$x + 0 = x \text{ and } \widetilde{0} + y = y.$$

Choosing $x = \widetilde{0}$ and $y = 0$ above, we obtain

$$\widetilde{0} = \widetilde{0} + 0 = 0,$$

showing the uniqueness of the additive identity.

Similarly, if 1 and $\widetilde{1}$ are two multiplicative identities, then for all $x, y \in \mathbb{F}$,

$$1 \times x = x \text{ and } y \times \widetilde{1} = y.$$

Substituting $x = \widetilde{1}$ and $y = 1$ into the above identities, we find that

$$\widetilde{1} = 1 \times \widetilde{1} = 1,$$

as claimed.

To prove the uniqueness of the additive inverse, let $x \in \mathbb{F}$, and let $-x \in \mathbb{F}$ be its additive inverse, so that

$$x + (-x) = 0.$$

If y is another additive inverse of x, then

$$x + y = 0. \tag{1.1}$$

Adding $-x$ to both sides of (1.1) and using the associativity of addition it follows that

$$-x = -x + 0 = -x + (x + y) = ((-x) + x) + y = 0 + y = y,$$

which proves the uniqueness.

The proof of claim 4 is analogous and is left as an exercise (Problem 7). The proofs of claims 5 and 6 are left as an exercise (Problems 8 and 9). □

The set of real numbers, in addition to being a field with respect to the operations of addition and multiplication, has the additional property of being ordered.

Definition 1.6. *The field $\mathbb{F}$ is an* ordered field *if there is an ordering $<$ on it that satisfies the following axioms:*

1. *For all $x, y \in \mathbb{F}$, one and only one of the following is true: $x = y$, $x < y$, or $y < x$.*

2. *If $x < y$ and either $z < w$ or $z = w$, then $x + z < y + w$.*

3. *If $0 < x$ and $0 < y$, then $0 < x \times y$.*

The axioms of ordering are satisfied by the field of real numbers $\mathbb{R}$, and they are what makes some of the familiar properties of real numbers true. Some of them are summarized in the following proposition. As usual, we use the notation $x > y$ if and only if $y < x$.

Theorem 1.7. *For $x, y, z \in \mathbb{R}$, the following properties hold:*

1. *If $x < y$ and $y < z$, then $x < z$.*
2. *If $x > 0$ and $y > 0$, then $x + y > 0$.*
3. *If $x > 0$, then $-x < 0$.*
4. *If $x \neq 0$, then either $x > 0$ or $-x > 0$.*
5. *If $x < y$, then $-x > -y$.*
6. *If $x \neq 0$, then $x \times x = x^2 > 0$.*
7. *If $0 < x < y$, then $x^{-1} > y^{-1}$.*

Proof.

1. By the second axiom of ordering,
$$x < y \text{ and } y < z \text{ imply } x + y < y + z.$$
Further, the second axiom allows us to add the same number $-y$ to both sides, implying that
$$(x + y) + (-y) < (y + z) + (-y).$$
By the commutativity and associativity of the addition, we arrive at
$$x = x + (y + (-y)) < (y + (-y)) + z = z,$$
which proves the claim.

2. This follows immediately from axiom 2.

3. By the second axiom,
$$x > 0 \text{ implies } x + (-x) > 0 + (-x),$$
from which it follows that $-x < 0$.

4. From the first ordering axiom, if $x \neq 0$, then either $x > 0$ or $x < 0$. If $x < 0$, from the second axiom it follows that
$$0 = x + (-x) < 0 + (-x) = -x,$$
as claimed.

5. From $x < y$ and axiom 2 we have that
$$0 = x + (-x) < y + (-x),$$
and applying axiom 2 again, we have
$$-y = -y + 0 < -y + y + (-x) = -x,$$
as claimed.

6. If $x > 0$, then ordering axiom 3 implies that $x \times x = x^2 > 0$.

   If $x < 0$, then, from part 3, $-x > 0$. Recalling that by claim 6 of Theorem 1.5, $-x = (-1) \times x$, the fact that multiplication is associative and commutative implies that

$$0 < (-x) \times (-x) = (-1) \times x \times (-1) \times x = \big((-1) \times (-1)\big) \times x^2.$$

   The claim follows, since by claim 6 of Theorem 1.5, $(-1) \times (-1) = -(-1) = 1$.

7. We begin by proving that if $x > 0$, then $x^{-1} > 0$. We proceed by contradiction. If $x^{-1} < 0$, then $-x^{-1} > 0$; hence, by axiom 3,

$$-1 = (-1) \times 1 = (-1) \times \underbrace{x^{-1} \times x}_{=1} = (-x^{-1}) \times x > 0,$$

   implying that $-1 > 0$, which contradicts part 3. Therefore

$$x > 0 \text{ implies that } x^{-1} > 0.$$

   Now, by the second axiom,

$$x < y \text{ implies } y + (-x) > x + (-x) = 0,$$

   and from ordering axiom 3 and the fact that $x^{-1} > 0$ it follows that

$$(y + (-1) \times x) \times x^{-1} > 0.$$

   Using the distributivity of the multiplication,

$$y \times x^{-1} + (-1) \times (x \times x^{-1}) > 0,$$

   or, equivalently,

$$y \times x^{-1} - 1 > 0.$$

   Since $y > 0$, we know that $y^{-1} > 0$, and we may use axiom 3 and the distributivity of the multiplication to obtain

$$y^{-1} \times (y \times x^{-1} - 1) = (y^{-1} \times y) \times x^{-1} - y^{-1} = x^{-1} - y^{-1} > 0,$$

   and then apply axiom 2 to prove the claim. □

Problems

1. Is the set of even integers a group with respect to the usual addition? Either prove that it satisfies all the properties for being a group or show that one of the properties is not satisfied. You can assume that 0 is an even number.

2. Is the set of odd integers a group with respect to the usual addition? Either prove that it satisfies all the properties for being a group or show that one of the properties is not satisfied. You can assume that 0 is an odd number.

3. Consider the functions
$$f(x) = \sin x, \qquad g(x) = x^2.$$
   (a) Determine the ranges of f and g.
   (b) Is $f \circ g$ well defined? If so, determine its range.
   (c) Is $g \circ f$ well defined? If so, determine its range.
   (d) If both $f \circ g$ and $g \circ f$ are well defined, are they equal? Justify your answer.

4. Consider the assignment
$$f : \mathbb{N} \to \mathbb{Z}, \qquad f(n) = z, \text{ where } z^2 = n.$$
   (a) Is f a function? Justify your answer.
   (b) If f is a function, is it injective?
   (c) If f is a function, is it surjective?

5. Is the function
$$f : \mathbb{R} \to \mathbb{R}, \quad f(x) = x^2,$$
injective? Surjective? Justify your answer.

6. Is the function
$$f : \mathbb{R} \to \mathbb{R}, \quad f(x) = \sin x,$$
injective? Surjective? Justify your answer.

7. Prove that in a field $\mathbb{F}$, if $x \in \mathbb{F}$, $x \neq 0$, its multiplicative inverse x^{-1} is unique and different from the additive identity.

8. Prove that in a field $\mathbb{F}$, for any $x \in \mathbb{F}$,
$$x \times 0 = 0 \times x = 0,$$
where 0 is the additive inverse.

9. Prove that in a field $\mathbb{F}$,
$$(-1) \times x = -x,$$
where -1 is the additive inverse of the multiplicative identity, and $-x$ is the additive inverse of x.

10. In the field of the rational numbers $\mathbb{Q}$, does the rule
$$\frac{m}{n} < \frac{p}{q} \Leftrightarrow m + n < p + q$$
define an ordering? Justify your answer.

Chapter 2
Complex Numbers

The set of complex numbers is often considered an extension of the set of real numbers. There are many instances in linear algebra where it is very natural to work with complex numbers, in particular in the context of eigenvalues and eigenvectors. Some of the operations among complex numbers are very similar to the corresponding operations among real numbers, while others are defined differently; however, the definitions coincide when restricted to the subset of real numbers. In this chapter we briefly review complex numbers and their properties.

2.1 ▪ Cartesian form of complex numbers

Complex numbers are an extension of real numbers, and are often represented as ordered pairs of real numbers, with the convention that the first coordinate is a real number, while the second coordinate is a pure imaginary number.

Geometrically, one can think of complex numbers as points in the xy-plane, in which case the real numbers are located on the x-axis, while another copy of $\mathbb{R}$, positioned along the y-axis, represents the imaginary part of the complex number. If we use the symbol i to indicate the imaginary part, we can write a complex number $z \in \mathbb{C}$ either as an ordered pair listing its real and imaginary coordinates, or as a sum of its real and imaginary parts:

$$z = (a, b) = a + ib.$$

We assume that the complex number i satisfies

$$i^2 = i \times i = -1.$$

We define the operation of addition on the set $\mathbb{C}$ of complex numbers according to the rule

$$(a + ib) + (c + id) = (a + c) + i(b + d),$$

or, if the complex numbers are represented as ordered pairs,

$$(a, b) + (c, d) = (a + c, b + d),$$

where the addition is the usual operation on real numbers.

The *complex conjugate* of the complex number $z = a + ib$ is $\overline{z} = a - ib$. Geometrically, the unary operation of complex conjugation corresponds to a reflection of the complex plane through the real axis. Obviously, complex conjugation maps each real number to itself.

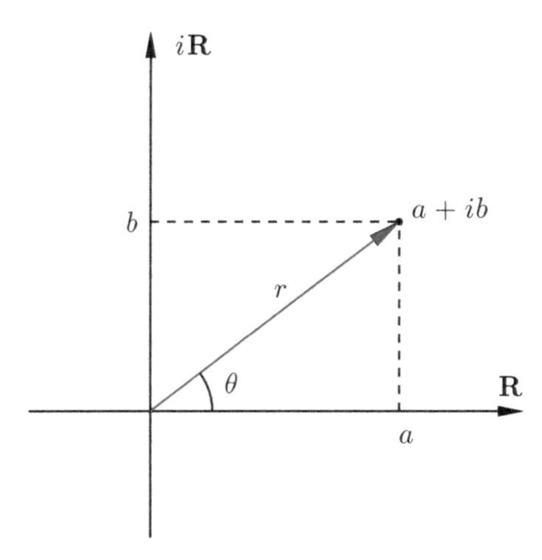

Figure 2.1. *Geometric interpretation of complex numbers as points in the complex plane.*

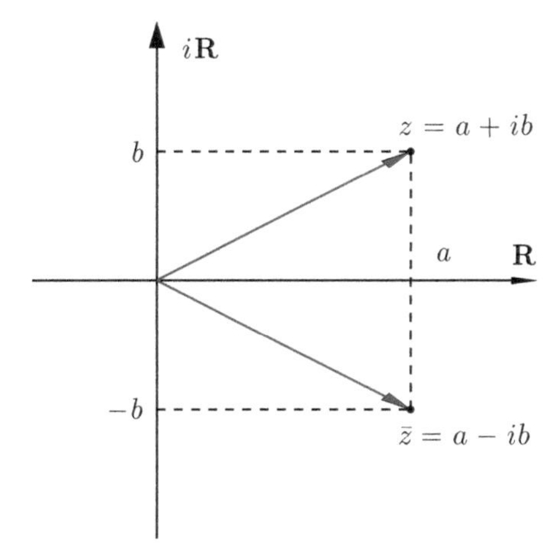

Figure 2.2. *Geometric interpretation of complex conjugation as a reflection through the real axis.*

Complex conjugation is distributive over addition:

$$\overline{z_1 + z_2} = \overline{z}_1 + \overline{z}_2.$$

Moreover, for every complex number z,

$$z + \overline{z} = 2\mathrm{Re}(z),$$

where $\mathrm{Re}(z)$ is the real part of z, that is, $\mathrm{Re}(a+ib) = a$. Similarly,

$$z - \overline{z} = 2i\,\mathrm{Im}(z),$$

where $\mathrm{Im}(z)$ is the imaginary part of z, defined as a real number, that is, $\mathrm{Im}(a+ib) = b$. Thus,

$$z = \mathrm{Re}(z) + i\,\mathrm{Im}(z) = \frac{z+\overline{z}}{2} + \frac{z-\overline{z}}{2}.$$

The operation of multiplication of two complex numbers is defined according to the rule

$$\begin{aligned}(a+ib)(c+id) &= ac+ibc+iad+i^2bd\\ &= ac-bd+i(bc+ad),\end{aligned}$$

where the usual notational convention $z_1 \times z_2 = z_1 z_2$ for multiplication is adopted. When multiplying together two complex numbers with zero imaginary part, their product as complex numbers is equal to their product as real numbers. Moreover, it is straightforward to verify that complex conjugation is distributive over multiplication, that is,

$$\overline{z_1 z_2} = \overline{z}_1 \overline{z}_2.$$

Theorem 2.1. *The set of complex numbers equipped with the operations of addition and multiplication defined above has the following properties:*

1. *Additive identity: The identity with respect to addition is the complex number*

$$0+i0,$$

compactly denoted by 0.

2. *Multiplicative identity: The identity with respect to multiplication is the complex number*

$$1+i0,$$

compactly denoted by 1.

3. *Additive inverse: For any* $a+ib \in \mathbb{C}$,

$$-(a+ib) = -a-ib.$$

4. *Multiplicative inverse: For any* $a+ib \neq 0$ *in* $\mathbb{C}$,

$$(a+ib)^{-1} = \frac{a}{a^2+b^2} - i\frac{b}{a^2+b^2}.$$

Proof. To prove 1, it suffices to verify that

$$(a+ib)+0+i0 = (a+0)+i(b+0) = a+ib,$$

while

$$(a+ib)(1+i0) = (a1-0b)+i(b1+a0) = a+ib$$

proves 2. The proof of 3 amounts to checking that the sum of a complex number and its presumptive additive inverse is 0.

The proof of 4 requires a little more work. Observe that a complex number is nonzero if at least one of its real and imaginary parts is nonvanishing; therefore

$$a+ib \neq 0 \Rightarrow a^2+b^2 > 0.$$

From the observation that

$$(a+ib)(a-ib) = aa+iba-iab-i^2bb = a^2+b^2,$$

it follows that

$$(a+ib)\frac{1}{a^2+b^2}(a-ib)=\frac{a^2+b^2}{a^2+b^2}=1;$$

hence

$$(a+ib)^{-1}=\frac{a}{a^2+b^2}-i\frac{b}{a^2+b^2},$$

which proves 4. □

Corollary 2.2. *The set of complex numbers $\mathbb{C}$ with the operations of addition and multiplication defined above is a field.*

The *modulus* of a complex number $z=a+ib$, defined as

$$|z|=\sqrt{z\bar{z}}=\sqrt{\mathrm{Re}(z)^2+\mathrm{Im}(z)^2}=\sqrt{a^2+b^2},$$

measures the Euclidian distance of the point $(\mathrm{Re}(z),\mathrm{Im}(z))$ of $\mathbb{R}^2$ from the origin.

2.2 ▪ Polar form

If $z=a+ib\neq 0$, then $a^2+b^2>0$; hence the point representing z in the complex plane lies on the circle centered at the origin of radius

$$r=\sqrt{a^2+b^2}. \tag{2.1}$$

Following up on this observation, we express a complex number in the form

$$z=a+ib=\sqrt{a^2+b^2}\left(\frac{a}{\sqrt{a^2+b^2}}+i\frac{b}{\sqrt{a^2+b^2}}\right),$$

where the point with coordinates$\left(\frac{a}{\sqrt{a^2+b^2}},\frac{b}{\sqrt{a^2+b^2}}\right)$ is on the circle of radius one centered at the origin. If θ is the angle between the positive real axis and the segment joining the origin to z, called the *phase*, then

$$\cos\theta=\frac{a}{\sqrt{a^2+b^2}},\qquad \sin\theta=\frac{b}{\sqrt{a^2+b^2}}, \tag{2.2}$$

and we can write

$$z=r\left(\cos\theta+i\sin\theta\right), \tag{2.3}$$

which is referred to as the *polar form* of the complex number z. A geometric interpretation of (2.3) is shown in Figure 2.1. Since

$$z=r\left(\cos\theta+i\sin\theta\right)=r\left(\cos(\theta+2k\pi)+i\sin(\theta+2k\pi)\right),\ k\in\mathbb{Z},$$

in order for the pair (r,θ) to be a unique set of coordinates of the complex number z, we need to restrict the range of θ, e.g., by requiring $0\leq\theta<2\pi$.

While the basic operations on complex numbers such as addition and multiplication are defined in terms of their Cartesian representations, others become much simpler when switching to the polar form.

2.3 ▪ Exponential form

There is a third way to represent complex numbers, besides the Cartesian and polar forms. Given Euler's formula for the complex exponential,

$$e^{ix} = \cos x + i \sin x,$$

where e is the base of the natural logarithm, every complex number can be expressed as

$$z = r\left(\cos\theta + i\sin\theta\right) = re^{i\theta}.$$

Complex exponentials follow the same product rules as real exponentials. It can be verified that the product of the two complex numbers

$$z_1 = r_1 e^{i\theta_1}, \quad z_2 = r_2 e^{i\theta_2}$$

is

$$z_1 z_2 = r_1 r_2 e^{i\theta_1} e^{i\theta_2} = r_1 r_2 e^{i(\theta_1+\theta_2)}. \tag{2.4}$$

The last equality follows from the formula for trigonometric functions of the sum of two angles. We have

$$\begin{aligned} e^{i\theta_1} e^{i\theta_2} &= (\cos\theta_1 + i\sin\theta_1)(\cos\theta_2 + i\sin\theta_2) \\ &= (\cos\theta_1\cos\theta_2 - \sin\theta_1\sin\theta_2) + i(\sin\theta_1\cos\theta_2 + \cos\theta_1\sin\theta_2) \\ &= \cos(\theta_1+\theta_2) + i\sin(\theta_1+\theta_2) \\ &= e^{i(\theta_1+\theta_2)}. \end{aligned}$$

The identity (2.4) turns out to be very convenient for calculating complex roots, as shown in the next section.

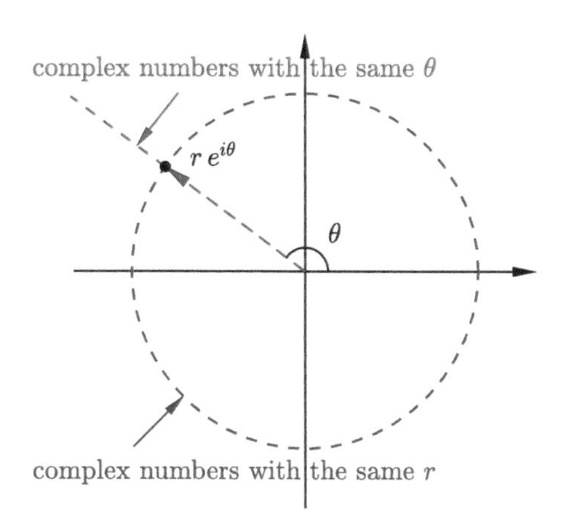

Figure 2.3. *Polar representation of complex numbers using the exponential form.*

2.4 ▪ De Moivre's formula and complex roots

The proof of the following theorem is an outstanding example of how important it is to choose the best suited representation of complex numbers for a given task.

Theorem 2.3 (De Moivre's formula). *If* $z = r\left(\cos\theta + i\sin\theta\right)$, *with* $r \geq 0$, *and* n *is a positive integer, then*

$$z^n = r^n\left(\cos n\theta + i\sin n\theta\right).$$

Proof. The proof is by induction on n. When $n = 1$, the statement holds trivially. Assume that the statement holds for $n-1$, that is,

$$z^{n-1} = r^{n-1}\left(\cos(n-1)\theta + i\sin(n-1)\theta\right),$$

which is the induction hypothesis. Next we want to show that it holds for n also. Using the exponential form, we write

$$z^{n-1} = r^{n-1}e^{i(n-1)\theta},$$

and by formula (2.4), we conclude that

$$\begin{aligned} z^n = zz^{n-1} &= \left(re^{i\theta}\right)\left(r^{n-1}e^{i(n-1)\theta}\right) = r^n e^{in\theta} \\ &= r^n\big(\cos n\theta + i\sin n\theta\big), \end{aligned}$$

completing the proof. □

The next theorem shows that, unlike in the field of real numbers, if k is a positive integer, every nonzero complex number has exactly k distinct kth roots. Furthermore, these roots are easy to characterize in polar representation.

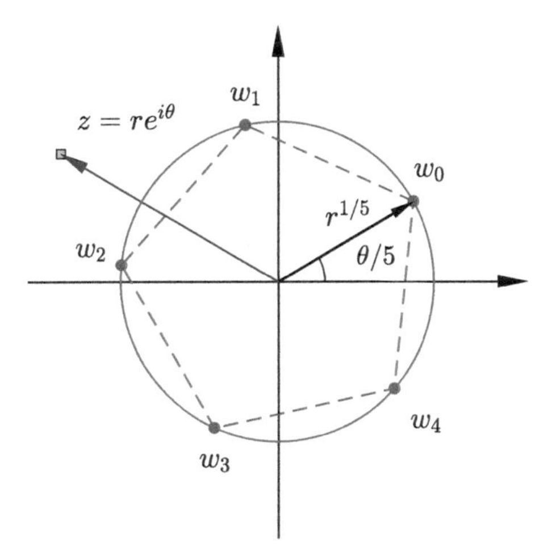

Figure 2.4. *Complex roots $w_0, \ldots, w_{k-1}$ of the equation $w^k = z$ with $k = 5$, where $z = re^{i\theta}$. The radius of the circle is $r^{1/5}$, and the roots are in the corners of a regular pentagon rotated by $\theta/5$.*

Theorem 2.4 (Roots of complex numbers). *If $z \neq 0$ is a complex number and $k \geq 0$ is an integer, there are exactly k distinct complex numbers w_j such that*

$$w_j^k = z, \quad 0 \leq j \leq k-1.$$

Geometrically, in the complex plane the k roots of z are represented by equispaced points on the circle centered at the origin and of radius $|z|^{1/k}$.

Proof. If

$$z = r\left(\cos\theta + i\sin\theta\right), \quad r = |z|,$$

then $w = \rho\left(\cos\varphi + i\sin\varphi\right)$ is a kth root of z if and only if $w^k = z$. It follows from de Moivre's formula that

$$\rho^k\left(\cos k\varphi + i\sin k\varphi\right) = r\left(\cos\theta + i\sin\theta\right). \tag{2.5}$$

The equality (2.5) holds if and only if $\rho = r^{1/k}$, and the phase angles satisfy

$$\cos k\varphi = \cos\theta,$$
$$\sin k\varphi = \sin\theta,$$

which, in turn, imply that

$$k\varphi = \theta + 2j\pi, \quad j \in \mathbb{Z},$$

or, equivalently,

$$\varphi = \frac{\theta}{k} + \frac{2j\pi}{k}.$$

Identifying complex numbers with the same modulus and phase angles differing by a multiple of 2π, we find k different solutions,

$$w_j = r^{1/k}\left(\cos\left(\frac{\theta}{k} + \frac{2j\pi}{k}\right) + i\sin\left(\frac{\theta}{k} + \frac{2j\pi}{k}\right)\right), \quad 0 \le j \le k-1. \qquad \square$$

Figure 2.4 shows the roots of the equation $w^5 = z$. The phase angles between the five roots are $2\pi/5$, placing the roots in the corners of a regular pentagon that is rotated counterclockwise by the angle $\theta/5$.

Problems

1. Find the multiplicative inverse of the complex number $z = -5 + 2i$ and write it in the form $c + id$. Then compute the inverse of z using its exponential representation.

2. Represent the two complex numbers $z_1 = 6 - 5i$ and $z_2 = -4 + 2i$ in the Cartesian plane as two vectors and verify graphically that their sum $w = z_1 + z_2$ as a vector is the sum of the two vectors z_1 and z_2.

3. Write in polar form $z = r(\cos\varphi + i\sin\varphi)$ the following complex numbers:

   (a) $z = -3i$,

   (b) $z = -2 - 8i$,

   (c) $z = -8$.

4. Given $z = -2 + 3i$ compute z^5 by

   (a) multiplying z by itself 5 times using its Cartesian representation;

   (b) using De Moivre's formula.

5. Compute the 4 distinct 4th roots of 81 and graph them in the Cartesian plane.

6. Compute the 3 distinct cubic roots of $-i$ and graph them in the Cartesian plane.

7. If $z = 5(\cos\theta + i\sin\theta)$, find $z\overline{z}$ and graph it in the Cartesian plane. Explain why you do not need to know what θ is.

8. Find all complex 5th roots of 1. Write them in polar form and plot them as points in the complex plane. How does the problem change if you look for all real roots instead?

9. Recall what it means that a complex number w is the 3rd root of z. Recalling that a monic polynomial can be written as the product of monomials $z - a_j$, where the a_j are its zeros, use the previous observation to factor the polynomial $p(z) = z^3 + 125$ as the product of monomials.

10. Find all 4th roots of $-16i$. Write them in polar form and plot them as points in the complex plane.

Chapter 3

Vector Spaces

Vector spaces are one of the pillars of linear algebra. Intuitively, a vector space is a generalization of the structure of $\mathbb{R}^2$ and $\mathbb{R}^3$ to more abstract spaces where the vectors do not have an immediate concrete geometric interpretation.

3.1 ▪ Definition and examples

We start by stating the formal definition of vector space.

Definition 3.1. *Let V be an Abelian group and $\mathbb{F}$ a field. We call V a* vector space over the field *$\mathbb{F}$ if it is closed with respect to the binary operation of* scalar multiplication

$$\mathbb{F} \times V \to V, \quad (\alpha, v) \mapsto \alpha v.$$

Moreover, the following properties must be satisfied:

1. *Associativity: $\beta(\alpha v) = (\beta\alpha)v$ for all $v \in V$, $\alpha, \beta \in \mathbb{F}$.*

2. *Distributivity: $(\beta + \alpha)v = \beta v + \alpha v$ for all $v \in V$, $\alpha, \beta \in \mathbb{F}$.*

3. *Scalar multiplication identity: $1v = v$ for all $v \in V$, where 1 is the multiplicative identity of $\mathbb{F}$.*

Many important sets with naturally defined operations of vector addition and scalar multiplication over a field are vector spaces. A set equipped with the operations of addition and scalar multiplication over a field is a vector space if all the properties of vector spaces are satisfied. Failure to satisfy any one of them disqualifies the set from being a vector space. In some cases, the obstacle to being a vector space arises from an unfortunate choice of the field $\mathbb{F}$; hence some sets may be vector spaces over one field but not over another.

Here are some examples of vector spaces that are routinely encountered in a variety of contexts.

Example 1: The set of all elements in

$$\mathbb{R}^n = \underbrace{\mathbb{R} \times \mathbb{R} \times \cdots \times \mathbb{R}}_{n}, \quad x = (x_1, x_2, \ldots, x_n)$$

for any integer $n \geq 1$, equipped with the componentwise addition,

$$(x_1, x_2, \ldots, x_n) + (y_1, y_2, \ldots, y_n) = (x_1 + y_1, x_2 + y_2, \ldots, x_n + y_n),$$

and scalar multiplication defined as

$$(\alpha, x) \mapsto (\alpha x_1, \alpha x_2, \ldots, \alpha x_n), \ \alpha \in \mathbb{R},$$

is a vector space over the field $\mathbb{F} = \mathbb{R}$.

Similarly, for any integer $n \geq 1$ the set of all vectors in $\mathbb{C}^n$ is a vector space over the fields of real or complex numbers, with the operations of addition and multiplication defined componentwise as in the case of $\mathbb{R}^n$. We remark that the choices of $\mathbb{F} = \mathbb{R}$ and $\mathbb{F} = \mathbb{C}$ are both possible, but lead to two different vector spaces.

In linear algebra, *matrices* are two-dimensional arrays with entries in a field. Each entry of a matrix is identified by two indices, the first referring to the row and the second to the column, as illustrated below:

$$\mathsf{A} = \begin{bmatrix} a_{11} & a_{12} & \cdots & a_{1n} \\ a_{21} & a_{22} & \cdots & a_{2n} \\ \vdots & \vdots & & \vdots \\ a_{m1} & a_{m2} & \cdots & a_{mn} \end{bmatrix} \in \mathbb{F}^{m \times n} \quad \text{if } a_{ij} \in \mathbb{F}.$$

Most commonly, $\mathbb{F} = \mathbb{R}$ or $\mathbb{F} = \mathbb{C}$, in which case A is an $m \times n$ real or complex matrix, respectively.

We define the sum of two matrices with the same number of rows and columns in a componentwise manner,

$$\begin{aligned} \mathsf{A} + \mathsf{B} &= \begin{bmatrix} a_{11} & a_{12} & \cdots & a_{1n} \\ a_{21} & a_{22} & \cdots & a_{2n} \\ \vdots & \vdots & & \vdots \\ a_{m1} & a_{m2} & \cdots & a_{mn} \end{bmatrix} + \begin{bmatrix} b_{11} & b_{12} & \cdots & b_{1n} \\ b_{21} & b_{22} & \cdots & b_{2n} \\ \vdots & \vdots & & \vdots \\ b_{m1} & a_{m2} & \cdots & b_{mn} \end{bmatrix} \\ &= \begin{bmatrix} a_{11} + b_{11} & a_{12} + b_{12} & \cdots & a_{1n} + b_{1n} \\ a_{21} + b_{21} & a_{22} + b_{22} & \cdots & a_{2n} + b_{2n} \\ \vdots & \vdots & & \vdots \\ a_{m1} + b_{m1} & a_{m2} + b_{m2} & \cdots & a_{mn} + b_{mn} \end{bmatrix}. \end{aligned}$$

The commutativity of matrix addition follows from the commutativity of the addition in the field $\mathbb{F}$. Scalar multiplication by elements of $\mathbb{F}$ is defined in a componentwise manner also:

$$\alpha \times \mathsf{A} \mapsto \begin{bmatrix} \alpha a_{11} & \alpha a_{12} & \cdots & \alpha a_{1n} \\ \alpha a_{21} & \alpha a_{22} & \cdots & \alpha a_{2n} \\ \vdots & \vdots & & \vdots \\ \alpha a_{m1} & \alpha a_{m2} & \cdots & \alpha a_{mn} \end{bmatrix}.$$

These observations lead to the following formal definition.

Definition 3.2. *The set of matrices, defined as arrays with m rows and n columns of entries in a field $\mathbb{F}$, form a vector space over the field $\mathbb{F}$.*

For later reference, a matrix of size $m \times n$ with zero entries is called a *null matrix*, denoted as $\mathsf{O}_{m \times n}$, or simply O, if the dimensionality is clear from the context.

Anticipating the notational conventions that will be adopted later on, we identify vectors in $\mathbb{R}^n$ with matrices in $\mathbb{R}^{n\times 1}$, representing vectors in $\mathbb{R}^n$ as column vectors, that is,

$$(x_1, x_2, \dots, x_n) \text{ is identified with } \begin{bmatrix} x_1 \\ x_2 \\ \vdots \\ x_n \end{bmatrix}.$$

Example 2: Let $I \subset \mathbb{R}$ denote an interval, and denote by $\mathcal{P}_k(I)$ the set of polynomials of $x \in I$ of degree at most $k \geq 0$ with real coefficients. In this case the vectors are the polynomials, and addition and scalar multiplication are defined as follows. If p and q are polynomials of degree at most k,

$$p(x) = a_0 + a_1x^1 + \cdots + a_kx^k, \quad q(x) = b_0 + b_1x^1 + \cdots + b_kx^k, \quad x \in I,$$

where $a_j, b_j \in \mathbb{R}$, and $\alpha \in \mathbb{R}$ is a scalar, we define

$$(p + \alpha q)(x) = (a_0 + \alpha b_0) + (a_1 + \alpha b_1)x^1 + \cdots + (a_k + \alpha b_k)x^k.$$

Observe that p and q can be polynomials of different orders, although not higher than k, since some of the coefficients may vanish.

3.2 ▪ Linear combinations, subspaces, and spanning sets

The sum of scalar multiples of vectors in a vector space V,

$$\alpha v + \beta w, \quad \alpha, \beta \in \mathbb{F}, \quad v, w \in V,$$

is called a *linear combination* of the vectors v and w. The definition of vector space implies that all linear combinations of vectors in a vector space over $\mathbb{F}$ with coefficients in $\mathbb{F}$ belong to the vector space.

Definition 3.3. *Let V be a vector space over the field $\mathbb{F}$. A subset $W \subset V$ is a* subspace *of V if it is a vector space with respect to the operations of addition and scalar multiplication defined on V. Equivalently, W is a subspace of V if it is* closed *under addition and scalar multiplication,*

1. *$v + w \in W$ for all $v, w \in W$,*
2. *$\alpha v \in W$ for all $\alpha \in \mathbb{F}$, $v \in W$.*

It follows directly from the definition of vector space that every subspace of a vector space must contain the additive identity of V, that is, $0 \in W$ is a necessary condition for W to be a subspace. Therefore if a subset of V does not contain $0 \in V$, it cannot be a subspace.

Conditions 1 and 2 above are equivalent to requiring that W contain the additive identity of V and

$$\alpha v + w \in W \text{ for all } \alpha \in \mathbb{F},\ v, w \in W.$$

Example 3: Consider the subset of vectors in $\mathbb{R}^2$ whose components (x_1, x_2) satisfy the equation of a line that does not pass through the origin, with the usual vector addition and scalar multiplication defined on $\mathbb{R}^2$. More specifically,

$$W = \{(x_1, x_2) \in \mathbb{R}^2 \mid (x_1, x_2) = (b_1, b_2) + t(v_1, v_2),\ t \in \mathbb{R}\},$$

where the vectors $b = (b_1, b_2) \neq 0$ and $v = (v_1, v_2) \neq 0$ are not collinear, that is,

$$b \neq \alpha v \text{ for all } \alpha \in \mathbb{R}.$$

Then W is not a subspace of $\mathbb{R}^2$ because it does not contain the additive identity. On the other hand, the subset of vectors in $\mathbb{R}^2$ identified with points on a line through the origin is a subspace of $\mathbb{R}^2$.

In a similar manner it can be shown that the subset of vectors in $\mathbb{R}^3$ corresponding to points in a plane not passing through the origin is not a subspace of $\mathbb{R}^3$, while the subset of vectors corresponding to points in a plane through the origin is a subspace of $\mathbb{R}^3$.

Definition 3.4. *A set of vectors $v_1, v_2, \ldots, v_k$ in a vector space V over the field $\mathbb{F}$ is a* spanning set *of V if every vector $v \in V$ can be expressed as their linear combination,*

$$v = \alpha_1 v_1 + \alpha_2 v_2 + \cdots + \alpha_k v_k, \quad \alpha_j \in \mathbb{F}.$$

A vector space may have many spanning sets, and the number of vectors in a spanning set may vary, as illustrated in the following example.

Example 4: The vectors

$$e_1 = \begin{bmatrix} 1 \\ 0 \\ 0 \end{bmatrix}, \quad e_2 = \begin{bmatrix} 0 \\ 1 \\ 0 \end{bmatrix}, \quad e_3 = \begin{bmatrix} 0 \\ 0 \\ 1 \end{bmatrix}$$

are a spanning set of $\mathbb{R}^3$ and are called the *canonical coordinate vectors*.

The set of vectors

$$\begin{bmatrix} 3 \\ 0 \\ 0 \end{bmatrix}, \quad \begin{bmatrix} 3 \\ 5 \\ 0 \end{bmatrix}, \quad \begin{bmatrix} 0 \\ 0 \\ 33 \end{bmatrix}, \quad \begin{bmatrix} 2 \\ 2 \\ 2 \end{bmatrix}$$

is also a spanning set of $\mathbb{R}^3$.

Example 5: Let $\mathcal{P}_k(I)$ be the space of polynomials of degree at most k in an interval $I \subset \mathbb{R}$ introduced in Example 3. The set of monomials

$$1, x, x^2, \ldots, x^k, \quad x \in I,$$

is a spanning set for $\mathcal{P}_k(I)$ since every polynomial of degree at most k can be expressed in the form

$$p(x) = a_0 + a_1 x + \cdots + a_k x^k.$$

The set of polynomials of degree less than or equal to k with leading coefficient one,

$$p_\ell(x) = \alpha_{0\ell} + \alpha_{1\ell} x + \cdots + x^\ell, \quad 0 \leq \ell \leq k,$$

is also a spanning set for the vector space.

Since, according to the definition, in order for a set of vectors to be a spanning set, each vector in the vector space can be obtained as their linear combination, the result stated in the following theorem is very natural.

Theorem 3.5. *Let V be a vector space over the field $\mathbb{F}$, and let $v_1, v_2, \ldots, v_k \in V$ be k given vectors. Then the set of all linear combinations of these vectors, denoted by*

$$\operatorname{span}\{v_1, \ldots, v_k\} = \left\{ \sum_{k=1}^{k} \alpha_j v_k, \mid \alpha_1, \ldots, \alpha_k \in \mathbb{F} \right\},$$

is a subspace of V spanned by the vectors $v_1, \ldots, v_k$. *This subset is called the* span *of $v_1, \ldots, v_k$.*

To prove the theorem it suffices to show that $\operatorname{span}\{v_1, \ldots, v_k\}$ is closed under addition and scalar multiplication.

3.3 ▪ Linear independence and bases

Intuitively, the linear independence of a set of vectors expresses the fact that each vector adds something new to the subspace spanned by the others. We begin by stating the formal definition of linear independence.

Definition 3.6. *The vectors $v_1, v_2, \ldots, v_n$ in the vector space V are* linearly independent *if their only linear combination that gives the zero vector is that with all zero coefficients, that is,*

$$\sum_{j=1}^{n} \alpha_j v_j = 0 \textit{ if and only if } \alpha_1 = \alpha_2 = \cdots = \alpha_n = 0. \tag{3.1}$$

The condition (3.1) in the definition will be used repeatedly to test the linear independence of a set of vectors in the following manner. If the vectors in the set are linearly independent, the *only* way to obtain the zero vector as their linear combination is if all coefficients vanish. Thus if there is any way to express the zero vector as a linear combination with at least one nonzero coefficient, the vectors cannot be linearly independent.

The following theorem sheds some light on the concept of linear independence, as well as on the origin of the term.

Theorem 3.7. *The vectors $v_1, v_2, \ldots, v_n$ in a vector space V are linearly independent if and only if it is not possible to express any one of them as a linear combination of the others.*

Proof. We start by proving by contradiction that *if* the vectors are linearly independent, we cannot express any one of them as a linear combination of the others. If the statement that we want to prove is false, then it is possible to express one of the vectors, for instance, v_1, as a linear combination of the others, that is, there exist scalars $\alpha_2, \ldots, \alpha_n$ such that

$$v_1 = \alpha_2 v_2 + \cdots + \alpha_n v_n.$$

In that case adding $-v_1 \in V$ to both sides yields

$$\begin{aligned} 0 &= -v_1 + \alpha_2 v_2 + \cdots + \alpha_n v_n \\ &= (-1)v_1 + \alpha_2 v_2 + \cdots + \alpha_n v_n. \end{aligned}$$

This is tantamount to saying that we can express the zero vector as a linear combination of the n vectors with coefficients that are not all vanishing, since the coefficient of v_1 is -1, contradicting the assumption that the vectors are linearly independent. Therefore the statement must be true.

We prove the converse, *only if* part, that is, that if none of the vectors can be expressed as a linear combination of the others, they are linearly independent, also by contradiction. If the vectors are not linearly independent, then there is a linear combination with coefficients not all vanishing yielding the zero vector, that is,

$$\alpha_1 v_1 + \alpha_2 v_2 + \cdots + \alpha_n v_n = 0.$$

Without loss of generality, we assume $\alpha_1 \neq 0$. Then

$$v_1 = -{\alpha_1}^{-1}\big(\alpha_2 v_2 + \cdots + \alpha_n v_n\big),$$

contradicting the assumption that we cannot represent one of the vectors as a linear combination of the others. The existence of α_1^{-1} follows from the existence of the multiplicative inverse of all nonzero elements in the field of scalars. □

The next proposition underlines the special role of the zero vector when it comes to linear independence.

Proposition 3.8. *The vectors in a set that includes the zero vector are not linearly independent.*

Proof. If the zero vector is in the set, it is always possible to express the zero vector as a linear combination of the vectors in the set with coefficients not all vanishing, by assigning a coefficient different from zero to the zero vector and setting all others to zero. Therefore the vectors are not linearly independent. □

Example 6: The vectors in $\mathbb{R}^3$,

$$v_1 = \begin{bmatrix} 9 \\ 0 \\ 0 \end{bmatrix}, \quad v_2 = \begin{bmatrix} 1 \\ 0 \\ 3 \end{bmatrix},$$

are linearly independent, because the only way to satisfy

$$\alpha_1 \begin{bmatrix} 9 \\ 0 \\ 0 \end{bmatrix} + \alpha_2 \begin{bmatrix} 1 \\ 0 \\ 3 \end{bmatrix} = \begin{bmatrix} 0 \\ 0 \\ 0 \end{bmatrix}$$

is to chose α_1 and α_2 so that

$$9\,\alpha_1 + \alpha_2 = 0, \qquad 3\,\alpha_2 = 0,$$

implying that

$$\alpha_1 = \alpha_2 = 0.$$

The vectors

$$w_1 = \begin{bmatrix} 2 \\ 4 \\ 6 \end{bmatrix}, \quad w_2 = \begin{bmatrix} 1 \\ 2 \\ 3 \end{bmatrix},$$

on the other hand, are not linearly independent because it can be shown that

$$\frac{1}{2} w_1 + (-1) w_2 = 0.$$

In this case the vector w_2 is a scaled version of w_1; hence w_2 adds nothing new to the subspace spanned by w_1.

We are now ready to formally define the basis of a vector space.

Definition 3.9. *A set of vectors is a* basis *of a vector space V if the vectors are*

1. *linearly independent,*
2. *a spanning set of V.*

The definition states that a basis of a vector space is a linearly independent spanning set. We remark that the existence of a basis of an arbitrary abstract vector space can be demonstrated, but the argument is nontrivial, relying on the Axiom of Choice. In general, it is not guaranteed that the basis is finite. In the discussion below, the existence of a finite basis is assumed.

The next theorem shows that once a basis for a vector space has been specified, every element of a vector space can be uniquely represented as a linear combination of the basis vectors.

Theorem 3.10. *If $v_1, v_2, \ldots$ is a basis of the vector space V, then for any $v \in V$,*

$$v = \sum_j \alpha_j v_j, \quad \alpha_j \in \mathbb{F},$$

for a unique choice of the scalars α_j. The scalars α_j are the coordinates *of v in the basis $v_1, v_2, \ldots$.*

Proof. Since the vectors in a basis are a spanning set for the vector space, for any $v \in V$ there exist scalars α_j such that

$$v = \sum_j \alpha_j v_j. \tag{3.2}$$

To prove the uniqueness of this representation, assume that there is a vector $v \in V$ that admits another representation as a linear combination of the basis vectors, e.g.,

$$v = \sum_j \beta_j v_j. \tag{3.3}$$

Subtracting the equations (3.3) and (3.2) side by side, it follows that

$$0 = \sum_j (\alpha_j - \beta_j) v_j.$$

The linear independence of the basis vectors implies that, for all j,

$$\alpha_j - \beta_j = 0, \text{ or } \alpha_j = \beta_j,$$

thus proving the uniqueness of the representation. □

The number of elements in a spanning set of a vector space with a finite dimensional basis may vary. The next theorem establishes an important relationship between the number of spanning vectors and the number of linearly independent vectors.

Theorem 3.11 (Exchange Theorem). *If V is a vector space spanned by a finite number $n \geq 1$ of vectors, there are at most n linearly independent vectors in the vector space.*

Proof. If $n = 1$, all vectors in V are scalar multiples of a vector v_1. Therefore if $w_1, w_2, \in V$, $w_1 \neq 0 \neq w_2$, then $w_1 = \alpha_1 v_1$, $w_2 = \alpha_2 v_1$ with $\alpha_1 \neq 0 \neq \alpha_2$ and $0 = w_1 - \frac{\alpha_1}{\alpha_2} w_2$, thus proving that w_1 and w_2 are not linearly independent.

We prove the statement for $n \geq 2$ by contradiction. Let

$$v_1, v_2, \ldots, v_n$$

be a spanning set for V. If the statement is not true, there are $m > n$ linearly independent vectors in V,

$$w_1, w_2, \ldots, w_m,$$

all different from zero. Since $v_1, v_2, \ldots, v_n$ constitute a spanning set, we can write

$$w_1 = \alpha_1 v_1 + \alpha_2 v_2 + \cdots + \alpha_n v_n, \quad \alpha_j \in \mathbb{F},$$

with coefficients α_j not all equal zero, because otherwise $w_1 = 0$, contradicting the linear independence of the w_js. Assuming, without loss of generality, that $\alpha_1 \neq 0$, we may write

$$v_1 = \alpha_1^{-1} \left(w_1 - \alpha_2 v_2 - \cdots - \alpha_n v_n \right),$$

which implies that

$$v_1 \in \operatorname{span} \left\{ w_1, v_2, \ldots, v_n \right\};$$

hence

$$V = \operatorname{span} \left\{ w_1, v_2, \ldots, v_n \right\}.$$

In this manner we have exchanged one element in the spanning set for an element in the linearly independent set, and any vector in V can be expressed as a linear combination of $w_1, v_2, \ldots, v_n$.

Next we express w_2 as a linear combination of $w_1, v_2, \ldots, v_n$, which are a spanning set for V,

$$w_2 = \beta_1 w_1 + \beta_2 v_2 + \cdots + \beta_n v_n, \quad \beta_j \in \mathbb{F}.$$

The coefficients β_j cannot be all zero; otherwise $w_2 = 0$, contradicting the linear independence of the w_j. Moreover, the coefficients of the v_j cannot all be zero because otherwise

$$w_2 = \beta_1 w_1 \;\Rightarrow\; \beta_1 w_1 - w_2 = 0,$$

once again contradicting the linear independence of the w_j. Without loss of generality, we may assume that $\beta_2 \neq 0$ and, proceeding as above, exchange v_2 for w_2 in the spanning set by observing that

$$v_2 = \beta_2^{-1} \left(w_2 - \beta_1 w_1 - \beta_3 v_3 - \cdots - \beta_n v_n \right);$$

hence

$$v_2 \in \operatorname{span}\{w_1, w_2, v_3, \ldots, v_n\}.$$

In general, if

$$V = \operatorname{span} \left\{ w_1, \ldots w_k, v_{k+1}, \ldots v_n \right\}, \; 1 \leq k \leq n-1, \tag{3.4}$$

then

$$w_{k+1} = \gamma_1 w_1 + \cdots + \gamma_k w_k + \gamma_{k+1} v_{k+1} + \cdots + \gamma_n v_n,$$

with the coefficients of the v_j not all equal to zero; otherwise w_{k+1} would be a linear combination of $w_1, \ldots, w_k$, contradicting the linear independence of the w_j. With reasoning analogous to that used for $k = 2$, it follows from (3.4) that

$$V = \text{span}\{w_1, w_2, \ldots, w_{k+1}, v_{k+2} \ldots v_n\}, \quad 1 \le k \le n-1,$$

and proceeding inductively, we arrive at the conclusion that

$$V = \text{span}\{w_1, \ldots, w_n\}.$$

If $m > n$, then $w_{n+1} \in V$ and there exist $\gamma_1, \ldots, \gamma_n$, not all equal to zero, such that

$$w_{n+1} = \gamma_1 w_1 + \ldots + \gamma_n w_n,$$

contradicting the linear independence of the w_j. Therefore, the number of linearly independent vectors cannot exceed the number of vectors in any spanning set, thus completing the proof. □

An important consequence of the Exchange Theorem is that if a vector space has a basis with a finite number n of vectors, then every basis has n vectors, as proved in the following corollary.

Corollary 3.12. *If the vector space V admits a finite basis, then every basis has the same number of vectors.*

Proof. Let

$$\mathcal{B}_1 = \{v_1, \ldots, v_n\}$$

and

$$\mathcal{B}_2 = \{w_1, \ldots, w_m\}$$

be two bases of V. Then the vectors in $\mathcal{B}_1$ and $\mathcal{B}_2$ are linearly independent and spanning sets. It follows from the Exchange Theorem that since the vectors in $\mathcal{B}_1$ are a spanning set for V and the vectors in $\mathcal{B}_2$ are linearly independent vectors, then $m \le n$. On the other hand, since the vectors in $\mathcal{B}_2$ are also a spanning set and those in $\mathcal{B}_1$ are linearly independent, then $n \le m$ also. This implies that $m = n$, completing the proof. □

If the number of elements in a basis of a vector space V is finite, by the previous corollary the number of the basis vectors is an invariant for the vector space, thus justifying the following definition.

Definition 3.13. *The dimension of a vector space V is the number of vectors in a basis.*

While the existence and finiteness of the basis are in general nontrivial questions, in concrete examples we can often construct the basis explicitly.

Example 7: It is straightforward to verify that the vectors

$$e_1 = \begin{bmatrix} 1 \\ 0 \end{bmatrix}, \quad e_2 = \begin{bmatrix} 0 \\ 1 \end{bmatrix}$$

are a basis for the vector space $\mathbb{R}^2$. Therefore every basis of $\mathbb{R}^2$ contains exactly two vectors.

In general, the vector space $\mathbb{R}^n$ has dimension n because the canonical coordinate vectors e_k, $1 \le k \le n$, with all components equal to zero and one in the kth position, are a basis.

When a vector in $\mathbb{R}^2$ is denoted by $\begin{bmatrix} x_1 \\ x_2 \end{bmatrix}$, the scalars x_1 and x_2 are its coordinates in the canonical basis $\begin{bmatrix} 1 \\ 0 \end{bmatrix}, \begin{bmatrix} 0 \\ 1 \end{bmatrix}$.

Example 8: In the vector space of polynomials with real coefficients of degree less than or equal to k, the monomials

$$1, x, x^2, \ldots, x^k$$

are a basis; hence every basis of the vector space must contain $k + 1$ polynomials.

Problems

1. Consider $\mathbb{R}^2$ with the usual addition of vectors. Let $\mathbb{F}$ be the field of real numbers and define the scalar multiplication as

$$(\alpha, x) = \begin{bmatrix} 2\alpha x_1 \\ -\alpha x_2 \end{bmatrix}.$$

   Is $\mathbb{R}^2$ with this scalar multiplication a vector space over $\mathbb{R}$? Justify your answer.

2. Consider the set W of all vectors in $\mathbb{R}^2$ whose coordinates $x,\ y$ satisfy the equation

$$y = 12x - 3.$$

   Is W a vector space? Justify your answer by verifying that either all the axioms of vector space are satisfied or at least one is not.

3. Is the set W of all vectors in $\mathbb{R}^2$ whose coordinates $x,\ y$ satisfy

$$y = 3x, \qquad -1 \leq x \leq 7,$$

   equipped with the componentwise addition and usual scalar multiplication by elements of $\mathbb{R}$ a subspace of $\mathbb{R}^2$? Justify your answer.

4. Is the set of all vectors in $\mathbb{R}^3$ whose coordinates $x,\ y,\ z$ satisfy the equation

$$2x - 5y + 3z = 0$$

   a vector space? Justify your answer.

5. Show that if V is a vector space over $\mathbb{F}$, for any $v \in V$ the scalar product of $0 \in \mathbb{F}$ times v is the additive identity in V, that is,

$$0_{\mathbb{F}} v = 0_V.$$

6. Show that if V is a vector space over $\mathbb{F}$, for any $v \in V$ the scalar product of $-1 \in \mathbb{F}$ times v is the additive inverse of v in V, that is,

$$(-1)v = -v.$$

7. Show that in a vector space V, the zero vector is always in $\operatorname{span}\{v_1, v_2, \ldots, v_k\}$, for $v_1, v_2, \ldots, v_k \in V$.

8. Let W be the set of vectors in $\mathbb{R}^4$ such that $x_2 = x_4 = 0$. Is W a subspace of $\mathbb{R}^4$?

9. Let W be the set of vectors in $\mathbb{R}^4$ such that $x_1 \geq x_4$. Is W a subspace of $\mathbb{R}^4$?

10. Let $z \in \mathbb{R}^4$ be a given vector, and let W be the set of vectors $w \in \mathbb{R}^4$ such that

$$\sum_{j=1}^{4} z_j w_j = 0,$$

   where z_j is the jth entry of z. Is W a subspace?

11. Let
$$W = \{w \in \mathbb{R}^4 \mid w_j \geq 0,\ 1 \leq j \leq 4\}.$$
Is W a subspace of $\mathbb{R}^4$?

12. Let $z, y \in \mathbb{R}^4$ be two given vectors and let
$$W = \left\{ w \in \mathbb{R}^4 \mid \sum_{j=1}^{4} w_j z_j = 0,\ \sum_{j=1}^{4} w_j y_j = 0 \right\}.$$
Is W a subspace of $\mathbb{R}^4$?

13. Let
$$W = \{w \in \mathbb{R}^4 \mid \cos w_4 = 1/2\}.$$
Is W a subspace of $\mathbb{R}^4$?

14. Is the set of vectors of the form
$$\begin{bmatrix} 3t + s \\ s + t \\ 2s \end{bmatrix}, \quad t, s \in \mathbb{R},$$
a subspace of $\mathbb{R}^3$? If so, find a basis for it.

15. Is the set of vectors of the form
$$\begin{bmatrix} 2t + s - r \\ s + t \\ 2s - r \\ 1 + t \end{bmatrix}, \quad t, s, r \in \mathbb{R},$$
a subspace of $\mathbb{R}^4$? If so, find a basis for it.

16. Are the following vectors a spanning set for $\mathbb{R}^4$? Justify your answer either by proving that every vector is $\mathbb{R}^4$ can be written as their linear combination or by finding at least one vector that cannot.
$$\begin{bmatrix} 1 \\ 1 \\ 1 \\ 0 \end{bmatrix}, \quad \begin{bmatrix} 0 \\ 0 \\ 1 \\ -1 \end{bmatrix}, \quad \begin{bmatrix} 0 \\ 3 \\ 0 \\ 0 \end{bmatrix}, \quad \begin{bmatrix} 0 \\ 0 \\ 1 \\ -2 \end{bmatrix}.$$

17. Are the vectors in the previous problem linearly independent? Justify your answer.

18. If you have 5 vectors in $\mathbb{R}^4$,

    (a) are they always linearly independent?

    (b) are they never linearly independent?

    (c) can they be linearly independent but not have to be?

    Remember to justify your answer.

19. If you have 2 vectors in $\mathbb{R}^3$,

    (a) are they always a spanning set for $\mathbb{R}^3$?

    (b) are they never a spanning set for $\mathbb{R}^3$?

    (c) are they always linearly independent?

    (d) are they never linearly independent?

    (e) are they guaranteed to be a basis for $\mathbb{R}^3$?

    Justify your answers.

20. Which of the sets of vectors

    (a) $\{x - 4x^3 - 2x^2 + 3x, x^2 - x, -x^3 + 25\}$, $x \in \mathbb{R}$,

    (b) $\{4 - 11x, 4x^2, -x^3 + 1, x^2 + x + 11\}$, $x \in \mathbb{R}$,

    are bases for the vector space of polynomials of degree less than or equal to 3? Justify your answers.

21. Which of the sets of vectors

    (a)
$$\begin{bmatrix} 9 \\ 10 \\ 0 \\ -2 \end{bmatrix}, \begin{bmatrix} 3 \\ 13 \\ 10 \\ -3 \end{bmatrix},$$

    (b)
$$\begin{bmatrix} -2 \\ 2 \\ 0 \\ 0 \end{bmatrix}, \begin{bmatrix} 0 \\ -1 \\ 0 \\ 1 \end{bmatrix}, \begin{bmatrix} 0 \\ 0 \\ 4 \\ 0 \end{bmatrix}, \begin{bmatrix} 2 \\ 0 \\ 0 \\ 0 \end{bmatrix},$$

    (c)
$$\begin{bmatrix} 2 \\ 10 \\ 0 \\ 1 \end{bmatrix}, \begin{bmatrix} 1 \\ 5 \\ 5 \\ 0 \end{bmatrix}, \begin{bmatrix} 7 \\ 4 \\ 5 \\ 0 \end{bmatrix}, \begin{bmatrix} -11 \\ 27 \\ -13 \\ -1 \end{bmatrix}, \begin{bmatrix} 0 \\ 0 \\ 8 \\ -6 \end{bmatrix}$$

    are bases for $\mathbb{R}^4$? Justify your answers.

22. Is the subset of $\mathbb{R}^3$ consisting of all vectors x with components x_1, x_2, and x_3 satisfying $4x_2 = x_1 + 2x_3$ a subspace of $\mathbb{R}^3$? If so, find a basis for it.

Chapter 4

Inner Products

The vector space property that the combination of vectors and scalars generates vectors also belonging to the vector space is what makes bases a very compact and convenient way of representing a vector space, similar to how two- and three-dimensional vectors in the Euclidean space can be expressed in terms of the canonical basis. A particularly useful property of the canonical basis is that the basis vectors are mutually perpendicular. In this chapter the geometric property of perpendicularity is extended to more general vector spaces.

In vector spaces over $\mathbb{R}$ or $\mathbb{C}$ it is often possible to define a function that maps ordered pairs of vectors to a real or complex number that is the natural generalization of the dot product. Given two vectors v and w in $\mathbb{R}^n$, their dot product is the sum of the product of their corresponding entries,

$$v = \begin{bmatrix} v_1 \\ \vdots \\ v_n \end{bmatrix}, \quad w = \begin{bmatrix} w_1 \\ \vdots \\ w_n \end{bmatrix}, \quad v \cdot w = \sum_{j=1}^{n} v_j w_j.$$

A slight change in the definition extends the dot products to vectors in $\mathbb{C}^n$. If $v, w \in \mathbb{C}^n$, let

$$v \cdot w = \sum_{j=1}^{n} v_j \overline{w_j},$$

where the bar denotes complex conjugation. In this manner, the dot product of two complex vectors with real entries coincides with the dot product of real vectors.

The most popular uses of dot products are to measure the size of a vector and to test whether two vectors are perpendicular to each other. In fact, it is easy to check that in $\mathbb{R}^2$ and $\mathbb{R}^3$,

$$v \perp w \;\Leftrightarrow\; v \cdot w = 0,$$

where the symbol $\perp$ was used to indicate the perpendicularity of the vectors.

We are now ready to formally define an analogue of the dot product on vector spaces over the field of the real or complex numbers.

Definition 4.1. *Let V be a vector space over the field $\mathbb{F} = \mathbb{R}$ or $\mathbb{C}$. The binary mapping*

$$\langle \cdot, \cdot \rangle : V \times V \to \mathbb{F}$$

is an inner product *on V if it satisfies the following properties for all $v, w, z \in V$ and $\alpha \in \mathbb{F}$:*

1. *Conjugate symmetry:* $\langle v, w\rangle = \overline{\langle w, v\rangle}$.

2. *Positivity:* $\langle v, v\rangle \geq 0$ *and* $\langle v, v\rangle = 0$ *if and only if* $v = 0$.

3. *Sesqulinearity:* $\langle \alpha v + w, z\rangle = \alpha\langle v, z\rangle + \langle w, z\rangle$.

A vector space equipped with an inner product is sometimes referred to as an *inner product space*.

Observe that the first and the third properties of inner products imply that

$$\langle v, \alpha w + z\rangle = \overline{\alpha}\langle v, w\rangle + \langle v, z\rangle.$$

In the case where $\mathbb{F} = \mathbb{R}$, the complex conjugate of a scalar is the scalar itself; thus the conjugate symmetry reduces to symmetry,

$$\langle v, w\rangle = \langle w, v\rangle,$$

and the sesquilinearity is referred to as *bilinearity*. In particular, when $\mathbb{F} = \mathbb{R}$, we have

$$\langle v, \alpha w + z\rangle = \alpha\langle v, w\rangle + \langle v, z\rangle.$$

It follows from the second property of inner products that the quantity $\langle v, v\rangle$ is a nonnegative real number, whose role is similar to that of the square of the length of a vector in Euclidean space.

In light of the previous observation, we define the length of a vector in a vector space equipped with an inner product to be

$$\|v\| = \sqrt{\langle v, v\rangle}.$$

It turns out that the length defined in this manner is a special instance of a norm, a concept that will be formally introduced and discussed in the next chapter. The dot product is the natural, or canonical, inner product for $\mathbb{R}^n$ and $\mathbb{C}^n$, and the length of vectors in terms of the dot product is the Euclidean length.

The dot product can be used to assess the relative orientation of two vectors in the Euclidean plane. Since the zero vector does not determine a direction in the plane, the following discussion is limited to the case of nonzero vectors. Two nonzero vectors v and w in $\mathbb{R}^2$ are orthogonal if they span a right angle, and in that case their dot product is zero. Since the angle between two nonzero vectors depends only on their mutual orientations in the plane, not on their lengths, it remains unchanged if we scale each vector by its length and perform a rigid rotation of the pair until the first vector v is along the positive direction of the x-axis and its second coordinate vanishes. If the two vectors are orthogonal, the second vector will lie along the y-axis; hence its x-coordinate vanishes and their dot product is zero. More generally, after rotating and scaling the two vectors v and w, we have

$$\frac{v}{\|v\|} = e_1, \quad \frac{w}{\|w\|} = \cos\varphi\, e_1 + \sin\varphi\, e_2,$$

where φ is the angle between the vectors, $0 \leq \varphi < \pi$. It follows that

$$\cos\varphi = \frac{v}{\|v\|} \cdot \frac{w}{\|w\|} = \frac{v \cdot w}{\|v\|\|w\|}.$$

Observe that for perpendicular vectors, $\varphi = \pi/2$, and $v \cdot w = 0$. This geometric observation gives a natural way to define orthogonality in vector spaces with an inner product.

4.1 ▪ Orthogonality

Because the inner product is a natural extension of the dot product, we define orthogonality of two vectors in terms of a given inner product as follows.

Definition 4.2. *Two vectors v and w in a vector space V equipped with an inner product are* orthogonal *if and only if*

$$\langle v, w \rangle = 0.$$

In this case we write $v \perp w$.

We remark that orthogonality is always relative to an inner product; hence we can think that the inner product induces some geometric properties on the underlying vector space. According to the definition of orthogonality, the zero vector is trivially orthogonal to all vectors with respect to any inner product. In fact, for any $v \in V$,

$$\langle 0_V, v \rangle = \langle 0_{\mathbb{F}} v, v \rangle = 0_{\mathbb{F}} \langle v, v \rangle = 0,$$

where the notations 0_V and $0_{\mathbb{F}}$ are used to emphasize that the former is the additive identity in the vector space, and the latter is the additive identity in the field of scalars.

Because the orthogonality of vectors in a vector space is always relative to an inner product, two vectors that are orthogonal with respect to an inner product are not necessarily orthogonal with respect to a different inner product.

The dot products in $\mathbb{R}^n$ and in $\mathbb{C}^n$ satisfy the conditions for being inner products; however, the class of inner products is much wider. For example, the assignment

$$\langle f, g \rangle = \int_0^1 f(x) g(x) dx \tag{4.1}$$

defines an inner product on the vector space $\mathcal{P}_k([0,1])$ of real-valued polynomials of degree up to k defined in the interval $[0,1]$. It is straightforward to verify that

$$\langle f, f \rangle = \int_0^1 f(x)^2 dx < \infty.$$

The natural generalization of this inner product to the space of polynomials of degree up to k with complex coefficients defined in the interval $[0,1]$ is the assignment

$$\langle f, g \rangle = \int_0^1 f(x) \overline{g(x)} dx.$$

A complex-valued function f is *square integrable* if

$$\langle f, f \rangle = \int_0^1 |f(x)|^2 dx < \infty.$$

It can be shown that the assignment (4.1), while defining an inner product for real-valued functions, is not an inner product on the space of complex-valued square integrable functions, failing to satisfy the first condition already.

The next theorem establishes an important relation between the modulus of the inner product of two vectors and the product of their lengths. In preparation, we prove two auxiliary results.

Lemma 4.3. *For any pair of vectors u, $v \neq 0$ in an inner product space, there exists a unique decomposition*

$$u = \alpha v + w, \tag{4.2}$$

where $w \perp v$, and $\alpha \in \mathbb{C}$ is given by

$$\alpha = \frac{\langle u, v \rangle}{\|v\|^2}. \tag{4.3}$$

Proof. Assume first that a decomposition of the form (4.2) exists. Then

$$\langle u, v \rangle = \langle \alpha v, v \rangle + \langle w, v \rangle = \alpha \langle v, v \rangle,$$

implying that the value of the scalar α is uniquely determined by (4.3), and necessarily,

$$w = u - \alpha v.$$

This demonstrates that if the decomposition exists, it is unique. On the other hand, if we let

$$u = \alpha v + (u - \alpha v),$$

where α is defined by (4.3), then

$$\langle u - \alpha v, v \rangle = \langle u, v \rangle - \frac{\langle u, v \rangle}{\|v\|^2} \langle v, v \rangle = 0,$$

thus proving the existence of (4.2). □

The following lemma is another way of stating Pythagoras' Theorem.

Lemma 4.4. *If*

$$u = v + w, \quad v \perp w,$$

then

$$\|u\|^2 = \|v\|^2 + \|w\|^2.$$

Proof. This is a straightforward calculation: expressing u in terms of v and w, we have that

$$\begin{aligned} \|u\|^2 &= \langle u, u \rangle = \langle v + w, v + w \rangle \\ &= \langle v, v \rangle + \langle v, w \rangle + \langle w, v \rangle + \langle w, w \rangle \\ &= \langle v, v \rangle + \langle w, w \rangle \\ &= \|v\|^2 + \|w\|^2, \end{aligned}$$

completing the proof. □

4.2 ▪ Cauchy–Schwarz inequality

We are now ready to prove one of the central results for inner products.

Theorem 4.5 (Cauchy–Schwarz inequality). *Let $\langle \cdot, \cdot \rangle$ be an inner product on a vector space V. Then for all $v, u \in V$,*

$$|\langle u, v \rangle| \leq \|u\| \|v\|, \tag{4.4}$$

with equality holding only if either

1. $v = 0$ *or* $u = 0$, *or*
2. $v \neq 0 \neq u$ *and* $u = \alpha v$, *for some* $\alpha \in \mathbb{F}$.

This result is known as the Cauchy–Schwarz inequality.

Proof. If $v = 0$ or $u = 0$, the claim holds with the equality sign.

Next we assume that $v \neq 0$. Using Lemma 4.3, we can write

$$u = \alpha v + w,$$

where α is given by (4.3). It follows from Lemma 4.4 that

$$\|u\|^2 = \|\alpha v\|^2 + \|w\|^2 \geq |\alpha|^2 \|v\|^2,$$

with equality holding if and only if $w = 0$, in which case $u = \alpha v$. Moreover, it follows from (4.3) that

$$\|u\|^2 \geq \frac{|\langle u, v\rangle|^2}{\|v\|^4}\|v\|^2 = \frac{|\langle u, v\rangle|^2}{\|v\|^2},$$

or, equivalently,

$$|\langle u, v\rangle|^2 \leq \|u\|^2 \|v\|^2,$$

thus completing the proof. □

We are now ready to generalize the definition of the angle between two vectors to the inner product vector space over the real field $\mathbb{R}$. The cosine of the angle φ between two vectors v, w, $v \neq 0 \neq w$, is defined according to the formula

$$\cos\varphi = \frac{\langle v, w\rangle}{\sqrt{\langle v, v\rangle}\sqrt{\langle w, w\rangle}} = \frac{\langle v, w\rangle}{\|v\|\|w\|}, \quad 0 \leq \varphi < \pi.$$

The Cauchy–Schwarz inequality guarantees that the quantity defined as the cosine of φ is in the interval $[-1, 1]$. In fact, if $v \neq 0$ and $w \neq 0$, the right-hand side of

$$|\langle v, w\rangle| \leq \|v\|\|w\|$$

is nonzero, and therefore

$$\frac{|\langle v, w\rangle|}{\|v\|\|w\|} \leq 1,$$

thus implying that the newly defined cosine satisfies

$$-1 \leq \cos\varphi \leq 1.$$

Sometimes, we use the notation $\varphi = \angle(v, w)$ to indicate the angle between vectors u and w.

4.3 ▪ Orthonormal bases

It follows from the properties of inner products that every vector $v \neq 0$ in an inner product vector space V can be scaled by the reciprocal of its length, or *normalized* to yield a vector of unit length.

Definition 4.6. *Let V be an n-dimensional inner product space. A basis $\{v_1, v_2, \ldots, v_n\}$ is an* orthonormal basis *if its elements are*

1. *mutually orthogonal,*
$$\langle v_j, v_k \rangle = 0 \textit{ for } j \neq k;$$

2. *normalized to have length 1,*
$$\|v_j\| = 1, \quad 1 \leq j \leq n.$$

The two conditions above are often written concisely in terms of the Kronecker symbol δ_{jk} as

$$\langle v_j, v_k \rangle = \delta_{jk} = \begin{cases} 1 & \text{if } j = k, \\ 0 & \text{if } j \neq k. \end{cases}$$

Orthogonality of vectors is a very strong property, in particular since it implies linear independence, as stated in the following theorem.

Theorem 4.7 (Orthogonality implies linear independence). *In the vector space V equipped with an inner product $\langle \cdot\, , \cdot \rangle$, any set of nonzero vectors*

$$v_1, v_2, \ldots, v_k$$

that are mutually orthogonal,

$$\langle v_i, v_j \rangle = 0, \; i \neq j,$$

are linearly independent.

Proof. Assume that there are scalars α_j such that

$$\sum_{j=1}^{k} \alpha_j v_j = 0. \tag{4.5}$$

To prove the linear independence of the vectors v_j, we need to show that (4.5) holds only if all coefficients vanish. Taking the inner product of both sides of (4.5) with v_ℓ yields

$$\left\langle \sum_{j=1}^{k} \alpha_j v_j, v_\ell \right\rangle = \sum_{j=1}^{k} \alpha_j \langle v_j, v_\ell \rangle = 0,$$

and since the inner products $\langle v_j, v_\ell \rangle$ all vanish except when $j = \ell$, we get

$$\alpha_\ell = 0.$$

This argument can be repeated for each ℓ, thus implying linear independence. □

This theorem is very powerful because orthogonality of given vectors is often easier to check than linear independence. The next result is very convenient when testing whether a set of vectors in a finite dimensional inner product vector space is a basis.

Theorem 4.8. *In a vector space V of dimension n equipped with an inner product, any n nonzero, mutually orthogonal vectors $w_1, w_2, \ldots, w_n$ are a basis.*

Proof. Since the linear independence of the orthogonal vectors is a corollary of the last theorem, it suffices to prove that every vector $v \in V$ can be written as a linear combination of the w_j. Consider the vectors $v, w_1, \ldots, w_n$. The n dimensionality of the vector space implies that the maximum number of linearly independent vectors is n; thus these $n+1$ vectors cannot be linearly independent. This implies the existence of scalars γ_j, not all vanishing, such that

$$0 = \gamma_0 v + \gamma_1 w_1 + \cdots + \gamma_n w_n. \tag{4.6}$$

It follows from the linear independence of the w_j that $\gamma_0 \neq 0$, because otherwise at least one of the coefficients $\gamma_1, \ldots, \gamma_n$ would be different from zero. If that were the case, (4.6) would be a linear combination of the vectors w_j with coefficients not all equal to zero yielding the zero vector. Solving (4.6) for v,

$$v = -\gamma_1/\gamma_0 w_1 - \cdots - \gamma_n/\gamma_0 w_n,$$

shows that v can be expressed as a linear combination of the w_j, thus proving that the vectors are a spanning set. □

In general, a way to assess how much two vectors overlap, or have a common component in them, is to take the orthogonal projection of one vector onto the other.

Definition 4.9. *In an inner product vector space V, the orthogonal projection of a vector w onto a vector $v \neq 0$ is the vector*

$$P_v(w) = \frac{\langle w, v\rangle}{\langle v, v\rangle} v = \left\langle w, \frac{v}{\sqrt{\langle v, v\rangle}} \right\rangle \frac{v}{\sqrt{\langle v, v\rangle}},$$

which is collinear with v. In the case of real vectors, the length of the vector depends on the cosine of the angle between w and v.

If w is orthogonal to v, we have

$$\langle w, v\rangle = 0 \Rightarrow P_v(w) = 0.$$

Observe that in general, by the Cauchy–Schwarz inequality, we have

$$\|P_v(w)\| = \frac{|\langle w, v\rangle| \|v\|}{\|v\|^2} \leq \frac{\|w\|\|v\|}{\|v\|} = \|w\|,$$

and the length of the projection of w onto v is largest when w is a multiple of v, e.g., $w = \alpha v$, in which case

$$P_v(w) = P_v(\alpha v) = \alpha \frac{\langle v, v\rangle}{\langle v, v\rangle} v = \alpha v = w,$$

that is, the projection is the vector w itself.

For each vector $v \neq 0$ in V, the assignment mapping each vector $w \in V$ to its projection onto v is a function called the *projector* onto v,

$$P_v : V \to V.$$

Since for every $w \in V$, $P_v(w)$ is a scalar multiple of v, we have

$$P_v P_v(w) = P_v(w) \text{ for all } w \in V.$$

Moreover, it follows from the linearity of the inner product that

$$P_v(\alpha_1 w_1 + \alpha_2 w_2) = \alpha_1 P_v(w_1) + \alpha_2 P_v(w_2).$$

Orthogonal projectors will be discussed in more generality later, in the context of solutions of linear systems in the least squares sense.

We close this chapter by showing that in an inner product space, from any set of linearly independent vectors, it is possible to produce a set of mutually orthogonal vectors of unit norm with analogous spanning properties. The proof of the following theorem is constructive and therefore yields an algorithm for building an orthonormal set.

Theorem 4.10 (Gram–Schmidt orthogonalization). *Let $v_1, v_2, \ldots, v_p$ be a set of linearly independent vectors in the inner product vector space V. Then there exist p orthonormal vectors $u_1, u_2, \ldots, u_p$ in V such that*

$$\text{span}\{v_1, v_2, \ldots, v_k\} = \text{span}\{u_1, u_2, \ldots, u_k\}, \quad 1 \leq k \leq p.$$

Proof. We prove the theorem by induction on the index $k < p - 1$ of the orthonormal vectors. We start by verifying that the statement holds for $k = 1$. In that case the vector $u_1 = v_1/\|v_1\|$ has unit norm, its span is the span of v_1, since it is a scalar multiple of v_1, and the orthonormality condition is trivially satisfied.

Next we assume that the statement holds for k, and we show that it holds for $k + 1$. The induction hypothesis states that there exists a set $\{u_1, u_2, \ldots, u_k\}$ of orthonormal vectors such that

$$\text{span}\{v_1, v_2, \ldots, v_k\} = \text{span}\{u_1, u_2, \ldots, u_k\}.$$

Consider the vector v_{k+1}: since the v_j are linearly independent,

$$v_{k+1} \notin \text{span}\{v_1, v_2, \ldots, v_k\},$$

and from the induction hypothesis, it follows that

$$v_{k+1} \notin \text{span}\{u_1, u_2, \ldots, u_k\}.$$

Therefore, we can define a nonvanishing vector z by the formula

$$z = v_{k+1} - \sum_{j=1}^{k} \alpha_j u_j, \tag{4.7}$$

where

$$\alpha_j = \langle v_{k+1}, u_j \rangle.$$

By construction,

$$z \in \text{span}\{u_1, u_2, \ldots, u_k, v_{k+1}\} = \text{span}\{v_1, v_2, \ldots, v_k, v_{k+1}\}. \tag{4.8}$$

Taking the inner product of z with u_ℓ, $1 \leq \ell \leq k$, from $\langle u_j, u_\ell \rangle = \delta_{j\ell}$, it follows that

$$\begin{aligned}\langle z, u_\ell \rangle &= \langle v_{k+1}, u_\ell \rangle - \sum_{j=1}^{k} \alpha_j \langle u_j, u_\ell \rangle \\ &= \langle v_{k+1}, u_\ell \rangle - \alpha_\ell = 0\end{aligned}$$

by definition of the coefficients α_j. This proves that z is orthogonal to every u_ℓ, $1 \leq \ell \leq k$; therefore the normalized vector

$$u_{k+1} = \frac{z}{\|z\|}$$

is orthogonal to all of the previous u_j. It follows from (4.7) that

$$v_{k+1} \in \operatorname{span}\{u_1, u_2, \ldots, u_k, z\} = \operatorname{span}\{u_1, u_2, \ldots, u_k, u_{k+1}\},$$

while (4.8) implies that

$$u_{k+1} \in \operatorname{span}\{v_1, v_2, \ldots, v_{k+1}\}.$$

We conclude that

$$\operatorname{span}\{v_1, v_2, \ldots, v_{k+1}\} = \operatorname{span}\{u_1, u_2, \ldots, u_{k+1}\},$$

which completes the induction proof. □

Problems

1. Prove that the assignment

$$\langle x, y\rangle = \sum_{j=1}^{3} x_j y_j$$

is an inner product in $\mathbb{R}^3$ but is not an inner product in $\mathbb{C}^3$.

2. Is the assignment

$$\langle x, y\rangle = \sum_{j=1}^{3} x_j (3y_j)$$

an inner product in $\mathbb{R}^3$? Justify your answer.

3. Consider the assignment

$$\langle \cdot, \cdot \rangle : \mathbb{C}^4 \times \mathbb{C}^4 \longrightarrow \mathbb{C}, \qquad \langle z, w\rangle = z_1 w_1 + 2z_2 w_2 + 3z_3 w_3 + 4z_4 w_4.$$

Is this an inner product in $\mathbb{C}^4$? Justify your answer.

4. Consider the assignment

$$\langle \cdot, \cdot \rangle : \mathbb{R}^4 \times \mathbb{R}^4 \longrightarrow \mathbb{R}, \qquad \langle x, y\rangle = x_1 y_1 + 2x_2 y_2 + 3x_3 y_3 + 4x_4 y_4.$$

Is this an inner product in $\mathbb{R}^4$? Justify your answer.

5. Verify that

$$\langle \cdot, \cdot \rangle : \mathbb{R}^3 \times \mathbb{R}^3 \longrightarrow \mathbb{R}, \qquad \langle x, y\rangle = 10x_1 y_1 + x_2 y_2 + 3x_3 y_3$$

is an inner product. With respect to this inner product, find the lengths of the vectors

$$\begin{bmatrix} 1 \\ -1 \\ 2 \end{bmatrix}, \quad \begin{bmatrix} 0 \\ 3 \\ -2 \end{bmatrix}$$

and the cosine of the angle between them. Are these two vectors orthogonal with respect to the given inner product?

6. If V is a vector space with an inner product $\langle \cdot, \cdot \rangle$, prove that for any $x, y \in V$,

$$\|x + y\|^2 + \|x - y\|^2 = 2\|x\|^2 + 2\|y\|^2,$$

where $\|v\|^2$ is the inner product of the vector v with itself. This is called the *parallelogram identity*. State at each step which property of inner products you are using. Give a geometric interpretation of the identity in $\mathbb{R}^2$.

7. If V is a vector space with an inner product $\langle \cdot, \cdot \rangle$, prove that for any $x, y \in V$,

$$\mathrm{Re}\langle x, y\rangle = \frac{1}{4}\left(\|x + y\|^2 - \|x - y\|^2\right),$$

where Re is the real part. This is called the *polarization identity*. State at each step which property of inner products you are using.

8. Consider the vector space $\mathcal{P}_4([-1,1])$ of the real-valued polynomials of degree less than or equal to 4 for $x \in [-1,1]$.

   (a) Verify that the assignment

   $$\langle p, q\rangle = \int_{-1}^{1} p(t)q(t)dt$$

   is an inner product.

   (b) With respect to the inner product defined above, compute the lengths of the polynomials

   $$p(t) = 2x^4 - x^2 - 1,$$
   $$q(t) = -x^3 - 4x,$$
   $$s(t) = 3x^2 - 3x + 1.$$

   (c) Are the polynomials $p(t)$ and $s(t)$ defined above orthogonal with respect to the inner product? Justify your answer.

9. Find the projection $P_{v_1}(v_2)$ of v_2 onto v_1. Then compute $z = v_2 - P_{v_1}(v_2)$ and the canonical inner product of z and v_1.

10. Find the projection $P_{v_1}(v_3)$ of v_3 onto v_1. Then compute $z = v_3 - P_{v_1}(v_3)$ and the canonical inner product of z and v_1.

11. Using Gram–Schmidt orthogonalization, find three orthonormal vectors u_1, u_2, u_3 such that

    $$\operatorname{span}\{u_1\} = \operatorname{span}\{v_1\}, \quad \operatorname{span}\{u_1, u_2\} = \operatorname{span}\{v_1, v_2\},$$
    $$\operatorname{span}\{u_1, u_2, u_3\} = \operatorname{span}\{v_1, v_2, v_3\}.$$

12. Show that if a vector z in $\mathbb{R}^3$ is orthogonal to the vectors $v_1 = (1,0,0)$, $v_2 = (0, \sqrt{2}/2, \sqrt{2}/2)$, then z, v_1, v_2 are a basis of $\mathbb{R}^3$.

Chapter 5

Vector Norms

In the Euclidean plane $\mathbb{R}^2$, the length of a vector, measured as the square root of the sum of the squares of its x and y coordinates, can be viewed as an application of Pythagoras' Theorem, where the vector is positioned along the hypotenuse of a suitable right triangle. The definition of length extends naturally to the Euclidean space $\mathbb{R}^3$, where it can be justified also in terms of Pythagoras' Theorem, and to $\mathbb{C}^2$ and $\mathbb{C}^3$ if we replace the square of the components with the square of their modulus.

In a general vector space, the analogue of Euclidean length as a means of assessing the size of a vector is a function from the vector space to the set of nonnegative real numbers that satisfies certain properties. This function is called a norm. Norms are important in linear algebra because they are the tools used for measuring the size of vectors and matrices, but their applicability extends way beyond linear algebra. In fact, norms play a key role in analysis and topology, making it possible, for example, to generalize the concept of distance to abstract spaces. We begin by stating the formal definition of norm.

5.1 ▪ Properties of norms

Definition 5.1. *In a vector space V over the field $\mathbb{F} = \mathbb{R}, \mathbb{C}$, the function*

$$\| \cdot \| : V \longrightarrow \mathbb{R}_+ = \{t \in \mathbb{R} \mid t \geq 0\}$$

is a norm *if it satisfies the following properties:*

1. *Nonnegativity: For all $v \in V$, $\|v\| \geq 0$ and $\|v\| = 0$ if and only if $v = 0$.*

2. *Scaling: For all $v \in V$ and $\alpha \in \mathbb{F}$, $\|\alpha v\| = |\alpha| \|v\|$.*

3. *Triangle inequality: For all $v, w \in V$,*

$$\|v + w\| \leq \|v\| + \|w\|.$$

A vector space equipped with a norm is called a normed vector space.

The scaling property is also referred to as *positive homogeneity* of the norm. Some of the most useful properties of norms follow directly from the definition. For example, the fact that for all $v \in V$,

$$\|-v\| = \|v\|$$

is an immediate corollary of the scaling property of norms, because

$$\|-v\| = \|(-1)v\| = |-1|\|v\| = 1\|v\| = \|v\|.$$

The triangle inequality implies that for all vectors v, w,

$$\big|\|v\| - \|w\|\big| \leq \|v - w\|,$$

which is seen as follows: If $\|v\| \geq \|w\|$, we have

$$\begin{aligned}
\big|\|v\| - \|w\|\big| &= \|v\| - \|w\| \\
&= \|v - w + w\| - \|w\| \\
&\leq \|v - w\| + \|w\| - \|w\| \\
&= \|v - w\|,
\end{aligned}$$

and similarly if $\|w\| > \|v\|$.

We remark that a norm is a function from a vector space to the nonnegative real numbers, regardless of whether the field of scalars of the vector space is $\mathbb{R}$ or $\mathbb{C}$.

To verify that a function from a vector space V to $\mathbb{R}_+$ is a norm it suffices to check that it satisfies the three properties in the definition. The natural norm in $\mathbb{R}^n$ is the Euclidean norm, also known as the 2-norm, denoted by $\|\cdot\|_2$, defined as the square root of the dot product of a vector with itself. For example, to show that the Euclidean length

$$\|x\|_2 = \left(\sum_{j=1}^{n} x_j^2\right)^{1/2}$$

is a norm in $\mathbb{R}^n$ we proceed as follows.

1. Since $x_j^2 \geq 0$ for all $x_j \in \mathbb{R}$, the sum on the right-hand side is always greater than or equal to zero. If one of the components of x, say x_k, is different from 0, then $x_k^2 > 0$; thus $\|x\|_2 > 0$. Therefore the only way to have $\|x\|_2 = 0$ is if all the components of x are zero, in which case $x = 0$.

2. To prove that the Euclidean length satisfies the second property of norms, notice that

$$\begin{aligned}
\|\alpha x\|_2^2 &= \sum_{j=1}^{n} (\alpha x_j)^2 \\
&= \alpha^2 \sum_{j=1}^{n} x_j^2 \\
&= \alpha^2 \|x\|_2^2.
\end{aligned}$$

Taking the square root of both sides completes the proof.

3. The prove the triangle inequality directly, observe that

$$\begin{aligned}\|x+y\|_2^2 &= \sum_{j=1}^{n}(x_j+y_j)^2\\ &= \sum_{j=1}^{n}\left(x_j^2+2x_jy_j+y_j^2\right)\\ &= \sum_{j=1}^{n}x_j^2+2\sum_{j=1}^{n}x_jy_j+\sum_{j=1}^{n}y_j^2\\ &= \|x\|_2^2+\|y\|_2^2+2\sum_{j=1}^{n}x_jy_j.\end{aligned}$$

On the other hand,

$$\begin{aligned}(\|x\|_2+\|y\|_2)^2 &= \|x\|_2^2+\|y\|_2^2+2\|x\|_2\|y\|_2\\ &= \|x\|_2^2+\|y\|_2^2+2\left(\sum_{j=1}^{n}x_j^2\right)^{1/2}\left(\sum_{j=1}^{n}y_j^2\right)^{1/2}.\end{aligned}$$

Thus, to complete the proof it suffices to show that

$$\sum_{j=1}^{n}x_jy_j \le \left(\sum_{j=1}^{n}x_j^2\right)^{1/2}\left(\sum_{j=1}^{n}y_j^2\right)^{1/2},$$

which is a consequence of the Cauchy–Schwarz inequality.

The next two norms are also used routinely for vectors in $\mathbb{R}^n$ and for vectors in $\mathbb{C}^n$. The 1-norm on $\mathbb{R}^n$ is defined as

$$\|x\|_1=\sum_{j=1}^{n}|x_j|=|x_1|+|x_2|+\cdots+|x_n|,$$

where $|x_j|$ is the modulus of x_j. The first property of norms is easy to verify because $|x_j|\ge 0$ for $1\le j\le n$; hence for any x, $\|x\|_1\ge 0$. Moreover, $\|x\|_1=0$ if and only if $|x_1|=|x_2|=\cdots=|x_n|=0$, or equivalently, $x=0$.

The scaling property is also straightforward to verify. For any $x\in\mathbb{C}^n$, $\alpha\in\mathbb{C}$,

$$\|\alpha x\|=\sum_{j=1}^{n}|\alpha x_j|=\sum_{j=1}^{n}|\alpha||x_j|=|\alpha|\sum_{j=1}^{n}|x_j|=|\alpha|\|x\|_1.$$

To prove that $\|\cdot\|_1$ satisfies the triangle inequality we observe that

$$\begin{aligned}\|x+y\|_1 &= \sum_{j=1}^{n}|x_j+y_j|\\ &\le \sum_{j=1}^{n}(|x_j|+|y_j|)\\ &= \sum_{j=1}^{n}|x_j|+\sum_{j=1}^{n}|y_j|\\ &= \|x\|_1+\|y\|_1.\end{aligned}$$

The ∞-norm on $\mathbb{C}^n$ is defined as

$$\|x\|_\infty = \max_{1\le j\le n} |x_j|.$$

For any $x \in \mathbb{C}^n$, $|x_j| \ge 0$; hence $\max_{1\le j\le n} |x_j| \ge 0$. Further, if $x = 0$, clearly $\|x\|_\infty = 0$. Conversely, if $\|x\|_\infty = 0$, then

$$\max_{1\le j\le n} |x_j| = 0,$$

implying that, for each j,

$$0 \le |x_j| \le \max_{1\le j\le n} |x_j| = 0,$$

from which we conclude that $x = 0$.

Next we observe that

$$\|\alpha x\|_\infty = \max_{1\le j\le n} |\alpha x_j| = |\alpha| \max_{1\le j\le n} |x_j| = |\alpha|\|x\|_\infty.$$

It remains to prove that $\|\cdot\|_\infty$ satisfies the triangle inequality. This follows from the observation that

$$\begin{aligned}
\|x+y\|_\infty &= \max_{1\le j\le n} |x_j + y_j| \\
&\le \max_{1\le j\le n} \big(|x_j| + |y_j|\big) \\
&\le \max_{1\le j\le n} |x_j| + \max_{1\le j\le n} |y_j| \\
&= \|x\|_\infty + \|y\|_\infty.
\end{aligned}$$

5.2 ▪ Norms induced by inner products

The 2-norm is a special instance of a norm that is defined in terms of an underlying inner product. The following theorem connects the previously defined length of a vector in an inner product space and the induced norm that assigns to each vector the square root of the inner product of the vector with itself.

Theorem 5.2. *In an inner product vector space V over $\mathbb{R}$ or $\mathbb{C}$, the function $\|\cdot\| : V \to \mathbb{R}^+$,*

$$\|v\| = \langle v, v\rangle^{1/2}$$

is a norm.

Proof. It suffices to verify that the three properties of norms are satisfied by the assignment.

1. Since $\langle\cdot\,,\cdot\rangle$ is an inner product,

$$\langle v, v\rangle \ge 0 \quad \text{and} \quad \langle v, v\rangle = 0 \text{ if and only if } v = 0;$$

hence

$$\|v\| \ge 0 \quad \text{and} \quad \|v\| = 0 \text{ if and only if } v = 0.$$

2. From the properties of inner products

$$\|\alpha x\| = \langle \alpha x, \alpha x\rangle^{1/2} = (\alpha\overline{\alpha})^{1/2}\langle x, x\rangle^{1/2} = |\alpha|\|x\|.$$

3. To prove the triangle inequality, notice that

$$\begin{aligned}\|v + w\|^2 &= \langle v + w, v + w\rangle \\ &= \langle v, v\rangle + 2\mathrm{Re}\langle v, w\rangle + \langle w, w\rangle \\ &\leq \|v\|^2 + 2|\langle v, w\rangle| + \|w\|^2.\end{aligned}$$

Because of its definition in terms of an inner product, we can use the Cauchy–Schwarz inequality,

$$|\langle v, w\rangle| \leq \|v\|\|w\|,$$

to show that

$$\begin{aligned}\|v + w\|^2 &\leq \|v\|^2 + 2\|v\|\|w\| + \|w\|^2 \\ &= (\|v\| + \|w\|)^2,\end{aligned}$$

as desired. □

An immediate consequence of Theorem 5.2 is that the triangle inequality is automatically satisfied by norms induced by inner products. Hence to prove that Euclidean norms satisfies the triangle inequality, it suffices to recall that they are induced by dot products, which are inner products.

If a candidate norm cannot be related to an inner product, it is necessary to check that it satisfies the three properties of norms. Of these, arguably, the triangle inequality is usually the more tedious one.

To assess how close to or far from each other vectors in V are, we need to introduce the notion of distance between pairs of vectors. The concept of distance is closely related to the idea of a metric. The properties of vector norms are closely related to the more general properties of a metric. While not a topic central to linear algebra, for completeness we state the definition of metric in a vector space.

Definition 5.3. *A vector space V is a* metric space *when equipped with a function*

$$d : V \times V \longrightarrow \mathbb{R}_+$$

called a metric *on V such that for all $x, y, z \in V$,*

1. $d(x, y) \geq 0,$ *and* $d(x, y) = 0$ *if and only if* $x = y$*;*

2. $d(x, y) = d(y, x)$*;*

3. $d(x, z) \leq d(x, y) + d(y, z)$.

We define the *distance* between two vectors v, w in normed vector space V as

$$d(v, w) = \|v - w\| = \|w - v\|.$$

It can be shown that the properties of norms imply that with this definition of distance, the vector space V is a metric space. Since the distance depends on the norm with respect to which it is computed, the distance between the same two vectors changes as we consider different norms. The dependency of the distance on the chosen norm makes it possible to change the way in which we assess how far apart or close together two vectors are, so as to best reflect the criteria that are used in the assessment.

In a normed vector space V, the *unit sphere* is the set of vectors of unit length,

$$S = \{v \in V : \|v\| = 1\}.$$

The unit spheres $\mathbb{R}^2$ and $\mathbb{R}^3$ are the sets of points at unit distance from the origin. To see how the unit sphere changes with the norm, in Figure 5.1 we show the unit spheres in $\mathbb{R}^2$ with respect to the 2-norm, 1-norm, and ∞-norm.

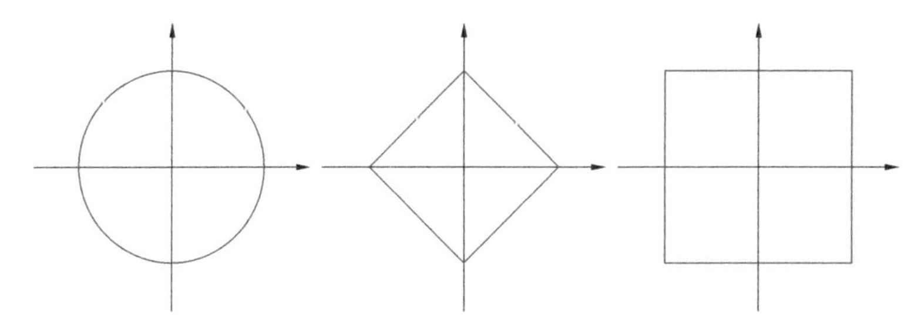

Figure 5.1. *The unit sphere in $\mathbb{R}^2$ in the 2-norm (left), 1-norm (center), and infinity norm (right).*

5.3 ▪ Equivalence of norms

Some vector norms are easier to evaluate numerically than others, and the difficulty of computing norms increases when considering norms of matrices. The following theorem establishes the equivalence of norms, up to a scaling factor, for all finite dimensional vector spaces over $\mathbb{C}$. The proof of the theorem makes use of the concept of continuity and some results from calculus.

Theorem 5.4 (Equivalence of norms). *Any two norms $\|\cdot\|$ and $|||\cdot|||$ over a finite dimensional vector space V over $\mathbb{C}$ are equivalent in the sense that there exist constants $0 < C_1 \le C_2$ such that, for any $x \in V$,*

$$C_1 |||x||| \le \|x\| \le C_2 |||x|||.$$

Proof. Let $\{v_1, \ldots, v_n\}$ be any basis of the vector space V, and define a third auxiliary norm, the ∞-norm related to this basis, as follows: If

$$x = \sum_{j=1}^{n} x_j v_j,$$

set

$$\|x\|_\infty = \max_{1 \le j \le n} \{|x_j|\}.$$

The vector space V equipped with this norm is denoted by V_∞. We start by showing that any given norm $\|\cdot\|$ is equivalent to $\|\cdot\|_\infty$. To this end, we show that the mapping

$$V_\infty \to \mathbb{R}, \quad x \mapsto \|x\| \tag{5.1}$$

is a continuous mapping. It follows from the triangle inequality that, for every $x, y \in V$,

$$\begin{aligned} \big| \|x\| - \|y\| \big| \le \|x - y\| &= \left\| \sum_{j=1}^{n} (x_j - y_j) v_j \right\| \\ &\le \sum_{j=1}^{n} |x_j - y_j| \|v_j\| \le \max_{1 \le j \le n} \{|x_j - y_j|\} \sum_{j=1}^{n} \|v_j\| \\ &= C\|x - y\|_\infty, \text{ where } C = \sum_{j=1}^{n} \|v_j\|. \end{aligned}$$

Therefore, for any $\varepsilon > 0$, $\big| \|x\| - \|y\| \big| < \varepsilon$ if $\|x - y\|_\infty < \delta = \varepsilon / C$, showing the continuity.

The unit sphere $S = \{x \in V \mid \|x\|_\infty = 1\}$ of V_∞ is closed and bounded, and it is a consequence of the Extreme Value Theorem that the continuous mapping (5.1) restricted to S attains its minimum and maximum values in S, denoted by

$$m = \min_{x \in S} \{\|x\|\}, \quad M = \max_{x \in S} \{\|x\|\}.$$

Notice that $m > 0$ because the norm is nonzero in the set S.

Let $x \in V$ be arbitrary, $x \neq 0$. Then

$$\widehat{x} = \frac{x}{\|x\|_\infty} \in S, \quad 0 < m \le \|\widehat{x}\| \le M,$$

and therefore, by multiplying the inequalities by $\|x\|_\infty$, we find that

$$0 < m\|x\|_\infty \le \|x\|_\infty \left\| \frac{x}{\|x\|_\infty} \right\| = \|x\| \le M\|x\|_\infty,$$

demonstrating the equivalence of the two norms $\| \cdot \|$ and $\| \cdot \|_\infty$.

To complete the proof, consider the vector norm $|||\cdot|||$. By the reasoning above, we can find constants m' and M' such that

$$0 < m'\|x\|_\infty \le |||x||| \le M'\|x\|_\infty.$$

Therefore,

$$\|x\| \le M\|x\|_\infty = \frac{M}{m'} m'\|x\|_\infty \le \frac{M}{m'} |||x|||,$$

and

$$\|x\| \ge m\|x\|_\infty = \frac{m}{M'} M'\|x\|_\infty \ge \frac{m}{M'} |||x|||,$$

or, equivalently,

$$c|||x||| \le \|x\| \le C|||x|||, \quad c = \frac{m}{M'}, \quad C = \frac{M}{m'},$$

which completes the proof. □

In concrete examples, it is often possible to find explicit expressions for constants C_1 and C_2 of the previous theorem, as the following example demonstrates.

Example 9: We show that in $\mathbb{R}^n$ and in $\mathbb{C}^n$,

$$\frac{1}{\sqrt{n}}\|x\|_1 \leq \|x\|_2 \leq \|x\|_1.$$

To prove the first inequality, we use the Cauchy–Schwarz inequality to deduce that

$$\begin{aligned}\|x\|_1 &= \sum_{j=1}^{n} |x_j| = \sum_{j=1}^{n} 1 \cdot |x_j| \\ &\leq \left(\sum_{j=1}^{n} 1^2\right)^{1/2} \left(\sum_{j=1}^{n} |x_j|^2\right)^{1/2} \\ &= \sqrt{n}\|x\|_2.\end{aligned}$$

To prove the second one, assume first that $\|x\|_1 = 1$. Then

$$\sum_{j=1}^{n} |x_j| = 1 \Rightarrow |x_j| \leq 1 \text{ for all } j,\ 1 \leq j \leq n;$$

therefore,

$$\begin{aligned}\|x\|_2^2 &= \sum_{j=1}^{2} |x_j|^2 = \sum_{j=1}^{2} \underbrace{|x_j|}_{\leq 1} \cdot |x_j| \\ &\leq \sum_{j=1}^{2} |x_j| = \|x\|_1 = 1,\end{aligned}$$

proving the claim for vectors with 1-norm equal to one. For a general $x \neq 0$, observe that the vector $x/\|x\|_1$ has a unit 1-norm, and therefore,

$$\left\| \frac{x}{\|x\|_1} \right\|_2 \leq 1.$$

The second inequality follows by the positive homogeneity of the norm. The constants $C_1 = 1/\sqrt{n}$ and $C_2 = 1$ are also the best possible ones in the sense that there are vectors for which these inequalities become equalities, meaning that the bounds cannot be improved. Indeed, by choosing

$$u = \begin{bmatrix} 1 \\ 1 \\ \vdots \\ 1 \end{bmatrix}, \quad v = \begin{bmatrix} 1 \\ 0 \\ \vdots \\ 0 \end{bmatrix},$$

we observe that

$$\frac{1}{\sqrt{n}}\|u\|_1 = \frac{n}{\sqrt{n}} = \sqrt{n} = \|u\|_2,$$

as is easy to check, while

$$\|v\|_2 = 1 = \|v\|_1.$$

It is left as an exercise to find the corresponding inequalities between the other pairs of norms in $\mathbb{C}^n$ (Problems 8 and 9).

Problems

Consider the vectors

$$v_1 = \begin{bmatrix} 1 \\ 0 \\ 2 \\ -4 \end{bmatrix}, \quad v_2 = \begin{bmatrix} 7 \\ 0 \\ -4 \\ -1 \end{bmatrix}, \quad v_3 = \begin{bmatrix} 5 \\ 2 \\ 0 \\ -6 \end{bmatrix}$$

in $\mathbb{R}^4$. Problems 1–7 refer to these vectors.

1. Compute the 1-norm of v_1, v_2, v_3.
2. Compute the 2-norm of v_1, v_2, v_3.
3. Compute the ∞-norm of v_1, v_2, v_3.
4. Without computing the inner product of v_1 and v_2, using what you computed above, can you find upper and lower bounds for their dot product? Justify your answer.
5. Show that for any vector v in $\mathbb{R}^n$, $\|v\|_\infty \le \|v\|_1$, and find a vector for which identity holds. What is the best constant C such that $\|v\|_1 \le C\|v\|_\infty$?
6. Find the best possible constants C_1 and C_2 such that

$$C_1\|x\|_\infty \le \|x\|_2 \le C_2\|x\|_\infty.$$

7. Explain why four vectors in $\mathbb{R}^3$ cannot be mutually orthogonal.

Chapter 6

Matrices

In mathematics, many concepts can be synthesized in the form of matrices, rectangular arrays of values in a field, organized into rows and columns. Matrices are representations of linear mappings between vector spaces, and matrices of a fixed size form a vector space themselves. In this section we introduce the definitions and results needed to work with matrices, and for analyzing them as self-contained objects or as linear operators.

6.1 ▪ Linear mappings, matrices, and vector spaces

We start with stating the formal definition of linear mapping.

Definition 6.1. *Let V and W be two vector spaces over the same field $\mathbb{F}$. A function $F : V \to W$ is a* linear map *if*

1. $F(\alpha u) = \alpha F(u)$ *for all* $\alpha \in \mathbb{F}$ *and* $u \in V$,
2. $F(u + v) = F(u) + F(v)$ *for all* $u, v \in V$.

The two conditions above can be combined in the following condition:

$$F(\alpha u + v) = \alpha F(u) + F(v);$$

choosing $v = 0$ gives the first condition, and $\alpha = 1$ yields the second.

It is common to denote the action of a linear operator omitting the parentheses when no risk of confusion exists, so we will identify

$$F(u) = Fu.$$

Let V and W be two finite dimensional vector spaces over the field $\mathbb{F}$, and let's assume that we can define them in terms of their respective bases as

$$V = \operatorname{span}\{v_1, v_2, \dots, v_n\}, \quad W = \operatorname{span}\{w_1, w_2, \dots, w_m\}.$$

By the properties of a basis, each vector Fv_j is a linear combination of vectors w_k; that is, there exist scalars $a_{kj} \in \mathbb{F}$ such that

$$Fv_j = \sum_{k=1}^{m} a_{kj} w_k, \quad 1 \le j \le n.$$

We organize the coefficients a_{kj} in a rectangular array and denote them by

$$\mathsf{A} = \begin{bmatrix} a_{11} & a_{12} & \cdots & a_{1n} \\ a_{21} & a_{22} & \cdots & a_{2n} \\ \vdots & & & \vdots \\ a_{m1} & a_{m2} & \cdots & a_{mn} \end{bmatrix} \in \mathbb{F}^{m\times n}. \tag{6.1}$$

Then A is the matrix representation of the linear mapping F in the bases $(\{v_j\}, \{w_k\})$. It is clear that the matrix of a linear map depends on the choice of the bases.

6.2 ▪ Basic operations and terminology

In this section we consider sets of matrices with different characteristics and introduce operations on matrices.

An $m \times n$ matrix with entries in $\mathbb{F} = \mathbb{R}, \mathbb{C}$ can be thought of as a collection of n vectors in $\mathbb{F}^m$, which are its columns, or, alternatively, as a collection of m vectors in $\mathbb{F}^n$ which are its rows.

The matrices of size $m \times n$ with entries in $\mathbb{F}$ form a vector space $\mathbb{F}^{m\times n}$ over the field $\mathbb{F}$, with addition defined as

$$\begin{bmatrix} a_{11} & \dots & a_{1n} \\ \vdots & & \vdots \\ a_{m1} & \dots & a_{mn} \end{bmatrix} + \begin{bmatrix} b_{11} & \dots & b_{1n} \\ \vdots & & \vdots \\ b_{m1} & \dots & b_{mn} \end{bmatrix} = \begin{bmatrix} a_{11}+b_{11} & \dots & a_{1n}+b_{1n} \\ \vdots & & \vdots \\ a_{m1}+b_{m1} & \dots & a_{mn}+b_{mn} \end{bmatrix}$$

and scalar multiplication as

$$\alpha \begin{bmatrix} a_{11} & \dots & a_{1n} \\ \vdots & & \vdots \\ a_{m1} & \dots & a_{mn} \end{bmatrix} = \begin{bmatrix} \alpha a_{11} & \dots & \alpha a_{1n} \\ \vdots & & \vdots \\ \alpha a_{m1} & \dots & \alpha a_{mn} \end{bmatrix}.$$

The zero of the vector space of matrices is the $m \times n$ matrix whose entries are all zeros, denoted by $\mathsf{O}_{m\times n}$, and the additive inverse of a matrix A is the matrix $-\mathsf{A}$ whose entries are, componentwise, the additive inverses of the entries of A.

It is straightforward to verify that a basis for the vector space $\mathbb{F}^{m\times n}$ consists of all matrices $\mathsf{B}_{ij} \in \mathbb{F}^{m\times n}$, $1 \le i \le m$, $1 \le j \le n$, whose entries are all zero except for the one in the (i, j) position. Therefore the vector space $\mathbb{F}^{m\times n}$ has dimension mn.

As long as we consider $m \times n$ matrices as arrangements of numbers, we can assemble their entries into mn vectors according to some agreed-upon rule and interpret $\mathbb{F}^{m\times n}$ as a different representation of $\mathbb{F}^{mn}$. Moreover, vectors $v \in \mathbb{F}^n$ can be regarded as matrices in $\mathbb{F}^{n\times 1}$, which will be a standard interpretation in what follows.

The *transpose* of an $m \times n$ matrix is an assignment $\mathbb{F}^{m\times n} \to \mathbb{F}^{n\times m}$, denoted by $\mathsf{A} \mapsto \mathsf{A}^\mathsf{T}$, resulting in a matrix whose rows are, respectively, the columns of A: The transpose of the matrix A in (6.1) is

$$\mathsf{A}^\mathsf{T} = \begin{bmatrix} a_{11} & a_{21} & \cdots & a_{m1} \\ a_{12} & a_{22} & \cdots & a_{m2} \\ \vdots & & & \vdots \\ a_{1n} & a_{2n} & \cdots & a_{mn} \end{bmatrix} \in \mathbb{F}^{n\times m}.$$

In particular, the transpose of a column vector in $v \in \mathbb{F}^m = \mathbb{F}^{m\times 1}$ is a row vector $v^\mathsf{T} \in \mathbb{F}^{1\times n}$ and vice versa.

In the case $\mathbb{F} = \mathbb{C}$, we define the *conjugate transpose*, or *Hermitian* of the $m \times n$ matrix A as an assignment $\mathbb{C}^{m\times n} \to \mathbb{C}^{n\times m}$, $\mathsf{A} \mapsto \mathsf{A}^{\mathsf{H}}$, where A^{H} is the matrix whose rows are the columns of A with the entries replaced by their complex conjugates,

$$\mathsf{A}^{\mathsf{H}} = \begin{bmatrix} \overline{a_{11}} & \overline{a_{21}} & \cdots & \overline{a_{m1}} \\ \overline{a_{12}} & \overline{a_{22}} & \cdots & \overline{a_{m2}} \\ \vdots & & & \vdots \\ \overline{a_{1n}} & \overline{a_{2n}} & \cdots & \overline{a_{mn}} \end{bmatrix} \in \mathbb{C}^{n\times m}.$$

.

A matrix A is *symmetric* if $\mathsf{A}^{\mathsf{T}} = \mathsf{A}$. A matrix A is *Hermitian*, or *self-adjoint*, if $\mathsf{A}^{\mathsf{H}} = \mathsf{A}$. These definitions imply that symmetric and Hermitian matrices must be square.

The operations of transposition and conjugate transposition coincide in the case of matrices with real entries, because the complex conjugate of a real number is the number itself, but are different operations for matrices with entries in $\mathbb{C}$.

6.3 ▪ Matrix-vector product

Consider an arbitrary vector $v \in V$, represented in terms of the basis $\{v_j\}$ as

$$v = \sum_{j=1}^{n} x_j v_j.$$

Let $F : V \to W$ be a linear mapping represented by the matrix A in the bases $\big(\{v_j\}, \{w_k\}\big)$, and assume that

$$Fv = w = \sum_{k=1}^{m} b_k w_k.$$

It follows from the linearity of the mapping that

$$\begin{aligned} Fv &= F\sum_{j=1}^{n} x_j v_j = \sum_{j=1}^{n} x_j F v_j \\ &= \sum_{j=1}^{n} x_j \sum_{k=1}^{m} a_{kj} w_k = \sum_{k=1}^{m} \left(\sum_{j=1}^{n} a_{kj} x_j \right) w_k. \end{aligned}$$

Since the representation of a vector in a given basis is unique, we conclude that

$$b_k = \sum_{j=1}^{n} a_{kj} x_j, \quad 1 \le k \le m.$$

This observation motivates the following definition of *matrix-vector product.*

Given $\mathsf{A} \in \mathbb{F}^{m\times n}$ and $x \in \mathbb{F}^{n\times 1} = \mathbb{F}^n$, we define the product of the matrix A and the vector x as

$$\mathsf{A}x = \begin{bmatrix} \sum_{j=1}^{n} a_{1j} x_j \\ \sum_{j=1}^{n} a_{2j} x_j \\ \cdots \\ \sum_{j=1}^{n} a_{mj} x_j \end{bmatrix} \in \mathbb{F}^{m\times 1} = \mathbb{F}^m.$$

Recalling that the columns of an $m \times n$ matrix can be regarded as vectors in $\mathbb{F}^m$, it is a straightforward matter to check that, in fact, $b = \mathsf{A}x$ is the linear combination of the columns of A scaled by the corresponding entries of the vector x,

$$\begin{bmatrix} b_1 \\ \vdots \\ b_m \end{bmatrix} = x_1 \begin{bmatrix} a_{11} \\ \vdots \\ a_{m1} \end{bmatrix} + x_2 \begin{bmatrix} a_{12} \\ \vdots \\ a_{m2} \end{bmatrix} + \cdots + x_n \begin{bmatrix} a_{1n} \\ \vdots \\ a_{mn} \end{bmatrix}.$$

With this definition in mind, it is clear that in order for the operation of matrix-vector product

$$\mathbb{F}^{m\times n} \times \mathbb{F}^n \longrightarrow \mathbb{F}^m, \quad (\mathsf{A}, x) \mapsto \mathsf{A}x$$

to be well defined, it is necessary that the number of columns of the matrix be equal to the number of elements in the vector. If this is the case, we say that the matrix and the vector are *compatible* with respect to the operation of matrix-vector product. The action of multiplying a vector in $\mathbb{F}^n$ by an $m \times n$ matrix implicitly defines a map from vectors in $\mathbb{F}^n$ to vectors in $\mathbb{F}^m$. Moreover, the assignment

$$\mathbb{F}^n \to \mathbb{F}^m, \quad x \mapsto \mathsf{A}x$$

is a linear mapping. Conversely, if A is the matrix representation of a linear map $F : U \to W$, then the associated matrix-vector product is a representation of this mapping in terms of the bases of the vector spaces U and V.

An important observation that will be elaborated on later is that the product of the matrix A and the vector x belongs to the subset of $\mathbb{F}^m$ of the linear combinations of the columns of the matrix A.

Furthermore, when $\mathbb{F} = \mathbb{R}$ or $\mathbb{C}$, the matrix-vector product allows an interpretation in terms of the natural inner product in $\mathbb{R}^n$ in the following way. Starting by regarding the vectors $v, w \in \mathbb{F}^n$ as matrices, the definition of the matrix-vector product

$$\begin{aligned} v^{\mathsf{T}} w &= \begin{bmatrix} v_1 & \cdots & v_n \end{bmatrix} \begin{bmatrix} w_1 \\ \vdots \\ w_n \end{bmatrix} \\ &= \sum_{j=1}^{n} v_j w_j = \langle v, w \rangle \end{aligned}$$

can be related to the canonical inner product in $\mathbb{R}^n$. We can now generalize this observation by partitioning a matrix $\mathsf{A} \in \mathbb{F}^{m\times n}$ row-wise, and think of its rows as transposes of m column vectors, $a_1^{\mathsf{T}}, \ldots, a_m^{\mathsf{T}}$, where $a_j \in \mathbb{F}^n = \mathbb{F}^{n\times 1}$. Then

$$\mathsf{A}x = \begin{bmatrix} \cdots & a_1^{\mathsf{T}} & \cdots \\ & & \\ \cdots & a_m^{\mathsf{T}} & \cdots \end{bmatrix} \begin{bmatrix} \vdots \\ x \\ \vdots \end{bmatrix} = \begin{bmatrix} \langle a_1, x \rangle \\ \vdots \\ \langle a_m, x \rangle \end{bmatrix}; \tag{6.2}$$

that is, we can compute the matrix-vector product $\mathsf{A}x$ by taking the canonical inner products in $\mathbb{R}^n$ of the vectors obtained by transposing the rows of A and the vector x. With this interpretation of the matrix-vector product, it is immediate to see that if a vector x is real orthogonal to the ℓth row of A, the ℓth entry of $\mathsf{A}x$ vanishes. More generally, if a vector x is real orthogonal to all rows of the matrix A, then $\mathsf{A}x = 0$. This observation will be revisited later.

6.4 ▪ Product of matrices

The operation of matrix-vector product can be extended to define matrix-matrix products.

Let $\mathbb{F} = \mathbb{R}$ or $\mathbb{C}$. Consider the matrices $\mathsf{A} \in \mathbb{F}^{m\times n}$, $\mathsf{B} \in \mathbb{F}^{n\times p}$, and regard B as the aggregate of its k column vectors, which are vectors in $\mathbb{F}^n$,

$$\mathsf{B} = \begin{bmatrix} \vdots & \vdots & & \vdots \\ b_1 & b_2 & \cdots & b_p \\ \vdots & \vdots & & \vdots \end{bmatrix}.$$

Define the matrix product of A and B as

$$\mathsf{AB} = \begin{bmatrix} \vdots & \vdots & & \vdots \\ \mathsf{A}b_1 & \mathsf{A}b_2 & \cdots & \mathsf{A}b_p \\ \vdots & \vdots & & \vdots \end{bmatrix} \in \mathbb{F}^{m\times p}.$$

Alternatively, if we regard A as the aggregate of its rows, from the interpretation (6.2) of matrix-vector product it follows that

$$\mathsf{AB} = \begin{bmatrix} \cdots & a_1^{\mathsf{T}} & \cdots \\ \cdots & a_2^{\mathsf{T}} & \cdots \\ & & \\ \cdots & a_m^{\mathsf{T}} & \cdots \end{bmatrix} \begin{bmatrix} \vdots & \vdots & & \vdots \\ b_1 & b_2 & \cdots & b_p \\ \vdots & \vdots & & \vdots \end{bmatrix} \tag{6.3}$$

$$= \begin{bmatrix} \langle a_1, b_1\rangle & \langle a_1, b_2\rangle & \cdots & \langle a_1, b_p\rangle \\ \langle a_2, b_1\rangle & \langle a_2, b_2\rangle & \cdots & \langle a_2, b_p\rangle \\ \vdots & & & \vdots \\ \langle a_m, b_1\rangle & \langle a_m, b_2\rangle & \cdots & \langle a_m, b_p\rangle \end{bmatrix},$$

where $\langle\ ,\ \rangle$ is the natural inner product in $\mathbb{R}^n$. Since the definition of matrix-matrix product is an extension of matrix-vector product, it is not surprising that the product of two matrices can be defined only between matrices that are *compatible*; that is, the number of columns of the first matrix must be equal to the number of rows in the second. The matrix-matrix product is a binary operation

$$\mathbb{F}^{m\times n} \times \mathbb{F}^{n\times p} \to \mathbb{F}^{m\times p}.$$

Matrix-matrix multiplication is associative, that is,

$$\mathsf{A(BC) = (AB)C},$$

when the matrices involved satisfy the compatibility condition: The number of rows in B must be equal to the number of columns in A, and the number of rows in C must be equal to the number of columns in B.

The matrix-matrix product is *not* commutative: the existence of the product AB does not even guarantee that the product of the matrices in reverse order is defined. In fact, if A is $m \times n$ and B is $n \times p$, in order for BA to be well defined it is necessary that $p = m$. In that case

$$\mathsf{AB} \in \mathbb{F}^{m\times m} \qquad \text{and} \qquad \mathsf{BA} \in \mathbb{F}^{n\times n};$$

therefore, if $m \neq n$, interchanging the order of the factors changes the dimensions of the results. If the two matrices A and B are both square of the same dimension, the products AB and BA both exist, but there is no guarantee that they are equal; in general, they are not.

If the product AB of the matrices A and B is defined, then the product of their transposes $\mathsf{B}^\mathsf{T}\mathsf{A}^\mathsf{T}$ is also defined, and

$$(\mathsf{AB})^\mathsf{T} = \mathsf{B}^\mathsf{T}\mathsf{A}^\mathsf{T}.$$

This holds also when transposition is replaced by complex conjugation, that is,

$$(\mathsf{AB})^\mathsf{H} = \mathsf{B}^\mathsf{H}\mathsf{A}^\mathsf{H}.$$

To prove this it suffices to show that the product in the right-hand side is well defined and that the expressions on both sides of the equal sign coincide, using, e.g., the interpretation (6.3).

6.5 ▪ Matrix inverse

The $n \times n$ matrix I_n whose entries are all zeros except for those on the main diagonal, which are all 1, is called the *identity matrix* of size $n \times n$,

$$\mathsf{I}_n = \begin{bmatrix} 1 & & \\ & \ddots & \\ & & 1 \end{bmatrix},$$

where we use the common notational convention of displaying zeros as empty spaces in the matrix. When there is no risk of confusion about the matrix dimensions, the identity matrix is sometimes written simply as I. The name refers to the fact that such a matrix acts as an identity for matrix multiplication: if $\mathsf{A} \in \mathbb{F}^{m\times n}$, where $\mathbb{F} = \mathbb{R},\ \mathbb{C}$, then $\mathsf{AI}_n = \mathsf{A}$, and if $\mathsf{B} \in \mathbb{F}^{n\times p}$, then $\mathsf{I}_n\mathsf{B} = \mathsf{B}$. In other words, multiplying a matrix by the identity matrix of compatible size leaves the matrix unchanged. When the dimension of the identity matrix is clear from the context, or of no consequence, we leave out the subindex and denote the identity matrix simply by I.

An $n \times m$ matrix B is a *left inverse* of the $m \times n$ matrix A if

$$\mathsf{BA} = \mathsf{I}_n,$$

and an $n \times m$ matrix C is a *right inverse* for A if

$$\mathsf{AC} = \mathsf{I}_m.$$

A *square matrix* $\mathsf{A} \in \mathbb{F}^{n\times n}$ is *invertible* if there is a matrix $\mathsf{B} \in \mathbb{F}^{n\times n}$ such that

$$\mathsf{AB} = \mathsf{BA} = \mathsf{I}_n. \tag{6.4}$$

The matrix B is called the inverse of A and is denoted by A^{-1}.

Theorem 6.2. *If a matrix is invertible, the inverse in unique.*

Proof. Let A^{-1} be the inverse of A, and let B be any matrix satisfying the conditions (6.4). Then

$$\mathsf{B} = \mathsf{BI} = \mathsf{B}\left(\mathsf{AA}^{-1}\right) = (\mathsf{BA})\mathsf{A}^{-1} = \mathsf{A}^{-1},$$

proving the uniqueness of the inverse. □

The next theorem proves that if a square matrix has both the left and the right inverses, then they coincide.

Theorem 6.3. *If a matrix $\mathsf{A} \in \mathbb{F}^{n\times n}$ has a left inverse B and a right inverse C, then A is invertible and*

$$\mathsf{B} = \mathsf{C} = \mathsf{A}^{-1}.$$

Proof. Since B is a left inverse and C is a right inverse of A, we have

$$\mathsf{BA} = \mathsf{I}, \qquad \mathsf{AC} = \mathsf{I},$$

implying that

$$\mathsf{B} = \mathsf{BI} = \mathsf{B(AC)} = \mathsf{(BA)C} = \mathsf{C}. \qquad \square$$

Theorem 6.4. *If a matrix is invertible, its conjugate transpose is also invertible and*

$$\left(\mathsf{A}^{\mathsf{H}}\right)^{-1} = \left(\mathsf{A}^{-1}\right)^{\mathsf{H}}.$$

Proof. To prove this theorem it suffices to verify that

$$\mathsf{A}^{\mathsf{H}}\left(\mathsf{A}^{-1}\right)^{\mathsf{H}} = \left(\mathsf{A}^{-1}\mathsf{A}\right)^{\mathsf{H}} = \mathsf{I}^{\mathsf{H}} = \mathsf{I},$$

and similarly,

$$\left(\mathsf{A}^{-1}\right)^{\mathsf{H}}\mathsf{A}^{\mathsf{H}} = \left(\mathsf{A}\mathsf{A}^{-1}\right)^{\mathsf{H}} = \mathsf{I}^{\mathsf{H}} = \mathsf{I}. \qquad \square$$

In the case where $\mathbb{F} = \mathbb{R}$ the conjugate of the matrix is the matrix itself; therefore the statement of the theorem becomes

$$\left(\mathsf{A}^{\mathsf{T}}\right)^{-1} = \left(\mathsf{A}^{-1}\right)^{\mathsf{T}}.$$

The theorem above justifies the shorthand notation

$$\mathsf{A}^{-\mathsf{H}} = \left(\mathsf{A}^{-1}\right)^{\mathsf{H}} = \left(\mathsf{A}^{\mathsf{H}}\right)^{-1},$$

which in the real case is written as $\mathsf{A}^{-\mathsf{T}}$.

The conditions guaranteeing that a matrix is invertible, and how to compute the inverses of some classes of matrices, will be addressed in detail later on. Furthermore, it can be proved with techniques discussed later in this book that if a square matrix has a left inverse, it also has the right inverse, and these coincide as shown in Theorem 6.3.

6.6 ▪ Orthogonal and unitary matrices

A matrix $\mathsf{Q} \in \mathbb{R}^{n\times n}$ is *orthogonal* if its columns, $q_1, \ldots, q_n \in \mathbb{R}^n$, are mutually orthogonal with respect to the standard inner product in $\mathbb{R}^n$ and are of unit norm, that is,

$$q_j^{\mathsf{T}} q_k = \langle q_j, q_k \rangle = \delta_{jk}, \quad 1 \le j, k \le n.$$

Therefore, if Q is orthogonal,

$$\mathsf{Q}^{\mathsf{T}}\mathsf{Q} = \mathsf{I} = \mathsf{Q}\mathsf{Q}^{\mathsf{T}}.$$

To show the first equality, we write the matrix-matrix product in terms of inner products of the row and column vectors,

$$\begin{bmatrix} \cdots & q_1^{\mathsf{T}} & \cdots \\ & & \\ \cdots & q_n^{\mathsf{T}} & \cdots \end{bmatrix} \begin{bmatrix} \vdots & & \vdots \\ q_1 & & q_n \\ \vdots & & \vdots \end{bmatrix} = \begin{bmatrix} q_1^{\mathsf{T}} q_1 & \cdots & q_1^{\mathsf{T}} q_n \\ \vdots & & \vdots \\ q_n^{\mathsf{T}} q_1 & \cdots & q_n^{\mathsf{T}} q_n \end{bmatrix} = \begin{bmatrix} 1 & & \\ & \ddots & \\ & & 1 \end{bmatrix}.$$

This identity shows that Q^{T} is a left inverse of Q. The second equality follows immediately from the remark at the end of the previous section and the fact that for a square matrix the left inverse is equal to the right inverse.

An immediate consequence of this result is that an orthogonal matrix is invertible and the inverse is its transpose,

$$\mathsf{Q}^{-1} = \mathsf{Q}^{\mathsf{T}}.$$

Moreover, the inverse of an orthogonal matrix is also an orthogonal matrix. In particular, the row vectors of an orthogonal matrix are mutually orthogonal.

A similar result holds for complex matrices. A matrix $\mathsf{U} \in \mathbb{C}^{n\times n}$ is *unitary* if its columns are mutually orthogonal with respect to the canonical inner product in $\mathbb{C}^n$ and have unit 2-norm. For a unitary matrix U,

$$\mathsf{U}^{\mathsf{H}}\mathsf{U} = \mathsf{I} = \mathsf{U}\mathsf{U}^{\mathsf{H}};$$

hence the inverse of a unitary matrix is its conjugate transpose,

$$\mathsf{U}^{-1} = \mathsf{U}^{\mathsf{H}}.$$

Orthogonal and unitary matrices have the property that they preserve the geometry of a vector space they operate on. More precisely, considering vectors with real entries, if $\mathsf{Q} \in \mathbb{R}^{n\times n}$ is an orthogonal matrix, or $\mathsf{U} \in \mathbb{C}^{n\times n}$ is a unitary matrix, then the following hold.

1. The 2-norm of a vector is unchanged by multiplication by Q or U:

$$\begin{aligned} \|x\|_2 &= \|\mathsf{Q}x\|_2 \text{ for all } x \in \mathbb{R}^n, \\ \|x\|_2 &= \|\mathsf{U}x\|_2 \text{ for all } x \in \mathbb{C}^n. \end{aligned}$$

2. The canonical inner product of two vectors in $\mathbb{R}^n$ or $\mathbb{C}^n$ is preserved after they are multiplied by Q or U, respectively:

$$\begin{aligned} \langle x, y\rangle &= \langle \mathsf{Q}x, \mathsf{Q}y\rangle \text{ for all } x, y \in \mathbb{R}^n, \\ \langle x, y\rangle &= \langle \mathsf{U}x, \mathsf{U}y\rangle \text{ for all } x, y \in \mathbb{C}^n. \end{aligned}$$

3. In particular, the angle between the two vectors is not changed by multiplication by Q:

$$\angle(x, y) = \angle(\mathsf{Q}x, \mathsf{Q}y) \text{ for all } x, y \in \mathbb{R}^n.$$

The proofs of these three statements follow directly from the definition of standard inner product expressed as matrix-matrix product. To prove 2, which implies 1 and 3, we observe that for an orthogonal matrix Q,

$$\langle \mathsf{Q}x, \mathsf{Q}y\rangle = (\mathsf{Q}x)^{\mathsf{T}}\mathsf{Q}y = x^{\mathsf{T}}\mathsf{Q}^{\mathsf{T}}\mathsf{Q}y = x^{\mathsf{T}}y = \langle x, y\rangle.$$

To prove 1, choose $x = y$; 3 follows from the definition of the angle between vectors.

Because of the conservation of the lengths and angles between vectors under multiplication by orthogonal or unitary matrices, these matrices correspond to rotations, reflections, and permutations of the components, or combinations thereof. In $\mathbb{R}^2$ or $\mathbb{R}^3$, these operations are easy to visualize as rotations around the origin and the reflections through a line or a plane passing through the origin, or swapping the coordinates. In higher dimensions, the generalization of these transformations can be visualized by limiting to lower dimensional projections, providing a geometric intuition when we discuss the connection between matrices and linear functions.

An important property of the set of orthogonal matrices is that it is closed with respect to matrix multiplication, as the following theorem states.

Theorem 6.5. *The product of orthogonal matrices is an orthogonal matrix. Similarly, the product of unitary matrices is a unitary matrix.*

Proof. If Q_1 and Q_2 are two orthogonal matrices, from $Q_1^T Q_1 = Q_1 Q_1^T = I$ and $Q_2^T Q_2 = Q_2 Q_2^T = I$ it follows that

$$\begin{aligned}(Q_1Q_2)(Q_1Q_2)^T &= Q_1Q_2Q_2^TQ_1^T \\ &= Q_1Q_1^T = I.\end{aligned}$$

The same argument is valid for the product in reverse order.

The proof for unitary matrices is essentially the same. □

To stress the importance of the fact that orthogonal matrices are square, consider the matrix $Q \in \mathbb{R}^{m\times n}$, $m \geq n$, with orthonormal columns; in this case $Q^TQ = I_n$; hence Q^T is a left inverse of Q. When $m > n$, Q^T is not a right inverse, because the n columns of Q are not a basis of $\mathbb{R}^m$.

6.7 ▪ Range of a matrix

Let $A \in \mathbb{F}^{m\times n}$ be a matrix with columns $a_1, \ldots, a_n \in \mathbb{F}^m$, where $\mathbb{F} = \mathbb{R},\ \mathbb{C}$. Given a vector $x \in \mathbb{F}^n$, the matrix-vector product is given as

$$Ax = x_1a_1 + x_2a_2 + \cdots + x_na_n,$$

a linear combination of the columns of A. Therefore, all vectors in $\mathbb{F}^m$ that can be obtained as products of A and vectors in $\mathbb{F}^n$ belong to the same subspace spanned by the columns of A, leading to the following definition.

Definition 6.6. *The set of vectors $\mathcal{R}(A) \subset \mathbb{F}^m$ that can be expressed as the product of the matrix A and a vector in $\mathbb{F}^n$ is a subspace of $\mathbb{F}^m$ called the* range *of the matrix A. The columns $a_1, a_2 \ldots, a_n$ are a spanning set of this subspace,*

$$\mathcal{R}(A) = \operatorname{span}\{a_1, a_2, \ldots, a_n\}.$$

If the columns of the matrix A are linearly independent, they are a basis of $\mathcal{R}(A)$ and the subspace is n-dimensional. In the case where the matrix $A \in \mathbb{F}^{m\times m}$ is a square matrix, the range of A is a subspace of $\mathbb{F}^m$. If the m columns of A are linearly independent, they are a basis for $\mathbb{F}^m$. Therefore every vector $y \in \mathbb{F}^m$ is a linear combination of the columns of A. Equivalently, there exists a vector $z \in \mathbb{F}^m$ such that $y = Az$. The entries of z are the coordinates of y in the basis of the columns of A, and to find them we need to solve the linear system

$$Az = y.$$

We therefore conclude that an $m\times m$ matrix A with linearly independent columns can be regarded as a change of coordinates in the space $\mathbb{F}^m$, because it maps the coordinates with respect to the basis formed by the columns of A to the canonical coordinates, or coordinates with respect to the canonical basis $\{e_1, \ldots, e_m\}$.

6.8 ▪ Null space of a matrix

Another useful subspace associated with a matrix is the null space.

Definition 6.7. *The set of all vectors in $\mathbb{F}^n$ that upon multiplication by an $m \times n$ matrix A are sent to the zero vector in $\mathbb{F}^m$ is a subspace of $\mathbb{F}^n$ called the* null space *of A and is denoted by*

$$\mathcal{N}(A) = \{x \in \mathbb{F}^n \mid Ax = 0\}.$$

The null space of a matrix always contains the zero vector of $\mathbb{F}^n$ since $\mathsf{A}0 = 0$. Therefore, to verify that the null space is indeed a subspace, it suffices to check that the subset is close under linear combinations.

If $x, z \in \mathcal{N}(\mathsf{A})$, then, by definition,

$$\mathsf{A}x = 0, \quad \mathsf{A}z = 0,$$

and for any $\alpha \in \mathbb{F}$,

$$\mathsf{A}\left(\alpha x + z\right) = \alpha \mathsf{A}x + \mathsf{A}z = 0$$

by the linearity of the mapping defined by matrix multiplication, thus proving that $\mathcal{N}(\mathsf{A})$ is a subspace of $\mathbb{F}^n$.

In the case where $\mathcal{N}(\mathsf{A}) = \{0\}$ we say that the matrix A has a trivial null space, and the dimension of the null space of A is zero. In general, the dimension of $\mathcal{N}(\mathsf{A})$ is less than or equal to n because $\mathcal{N}(\mathsf{A}) \subset \mathbb{F}^n$.

6.9 • Triangular orthogonalization: The QR factorization

At the end of section 6.7, we pointed out that if the columns of the square matrix A are linearly independent, $\mathsf{A}z = x$ can be seen as change of coordinates. In this section, we start working more systematically with the notion of change of basis. For clarity, we limit the discussion to vectors and matrices over a real field, although the analysis holds for complex matrices also.

The fact that orthogonal matrices preserve the geometry of the vector space of the vectors they multiply makes them play a special role among coordinate transformations. If $\mathsf{Q} \in \mathbb{R}^{n\times n}$ is an orthogonal matrix, the vectors $\mathsf{Q}e_j$, where e_j is the jth canonical basis vector, $1 \le j \le n$, are mutually orthogonal, thus linearly independent, and a basis for $\mathbb{R}^n$. The new basis vectors are the columns of the matrix Q. The coordinates of a vector $x \in \mathbb{R}^n$ in the new basis are the coefficients of the linear combination of the columns of Q that produces x, or, equivalently, the entries of the vector $z \in \mathbb{R}^n$ such that

$$\mathsf{Q}z = x.$$

Exploiting the fact that the inverse of an orthogonal matrix is its transpose, it is immediate to check that

$$z = \mathsf{Q}^{\mathsf{T}}x,$$

making the computation of the new coordinates simple and convenient. But what if the new basis defined by the columns of a matrix $\mathsf{A} \in \mathbb{R}^{n\times n}$ is not orthonormal? Theorem 4.10 provides a clue how to proceed.

Let $\{a_1, a_2, \ldots, a_n\}$ denote the columns of the matrix $\mathsf{A} \in \mathbb{R}^{n\times n}$, which we assume to be linearly independent. Theorem 4.10 guarantees the existence of an orthonormal basis $\{q_1, q_2, \ldots, q_n\}$ such that, for each k, $1 \le k \le n$,

$$\mathrm{span}\{a_1, \ldots, a_k\} = \mathrm{span}\{q_1, \ldots, q_k\}.$$

The construction of such a basis can be extracted from the inductive proof of the theorem. In particular, this means that for each k, $a_k \in \mathrm{span}\{q_1, \ldots, q_k\}$, implying that we can find coefficients r_{kj}, $1 \le j \le k$, such that

$$a_k = \sum_{j=1}^{k} r_{jk} q_j, \quad 1 \le k \le n. \tag{6.5}$$

In matrix-vector notation, we have

$$\begin{bmatrix} \vdots \\ a_k \\ \vdots \end{bmatrix} = \begin{bmatrix} \vdots & \vdots & & \vdots \\ q_1 & q_2 & \cdots & q_n \\ \vdots & \vdots & & \vdots \end{bmatrix} \begin{bmatrix} r_{1k} \\ r_{2k} \\ \vdots \\ r_{kk} \\ 0 \\ \vdots \\ 0 \end{bmatrix},$$

and expressing each column of A in the form (6.5) , we have

$$\mathsf{A} = \mathsf{QR}, \tag{6.6}$$

where $\mathsf{R} \in \mathbb{R}^{n\times n}$ is an *upper triangular matrix*,

$$\mathsf{R} = \begin{bmatrix} r_{11} & r_{12} & \cdots & r_{1n} \\ & r_{22} & & r_{2n} \\ & & \ddots & \vdots \\ & & & r_{nn} \end{bmatrix}.$$

Formula (6.6) is a particular instance of a more general matrix factorization result, called the *QR-factorization*, that is valid also for nonsquare matrices.

Theorem 6.8 (QR-factorization). *Every matrix $\mathsf{A} \in \mathbb{R}^{m\times n}$ admits a factorization*

$$\mathsf{A} = \mathsf{QR},$$

where $\mathsf{Q} \in \mathbb{R}^{m\times m}$ is an orthogonal matrix, and $\mathsf{R} \in \mathbb{R}^{m\times n}$ is upper triangular matrix; that is, the entries r_{jk} of R satisfy $r_{jk} = 0$, if $j > k$.

The proof of the theorem could be built on a process similar to the Gram–Schmidt orthogonalization. Alternatively, we can multiply A from the left by a sequence of orthogonal matrices designed so as to introduce zeros below the main diagonal. One way to design such a sequence of orthogonal matrices is to define an orthogonal matrix $\mathsf{H} \in \mathbb{R}^{m\times m}$ that maps a given vector x to a scalar multiple of the first canonical basis vector in $\mathbb{R}^m$, i.e., $\mathsf{H}x = \pm\|x\|e_1$.

One way to determine the matrix H is by following a geometric approach. For the time being, let us restrict our attention to the two-dimensional subspace spanned by the vectors x and $\|x\|e_1$, as illustrated in Figure 6.1. The vector

$$v = x - \|x\|e_1$$

is such that $\|x\|e_1 = x - v$. The vector to be added to x in order to obtain $\|x\|e_1$ is minus twice the projection of x onto v. Recall that the projection of x to v is

$$P_v(x) = \frac{v^\mathsf{T} x}{v^\mathsf{T} v} v;$$

therefore,

$$\begin{aligned} \|x\|e_1 &= x - v = x - 2P_v(x) \\ &= x - 2\frac{v^\mathsf{T} x}{v^\mathsf{T} v} v \\ &= x - 2\frac{vv^\mathsf{T}}{v^\mathsf{T} v} x, \end{aligned}$$

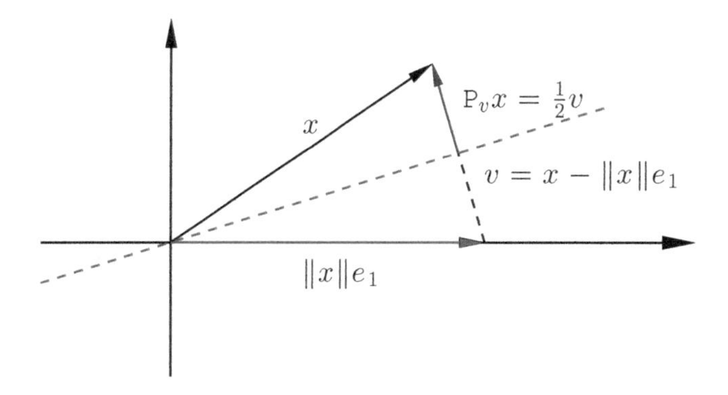

Figure 6.1. *A schematic description of the action of a Householder reflector.*

where we rearranged the products in the last equation:

$$\underbrace{(v^\mathsf{T} x)}_{\in \mathbb{R}} v = v(v^\mathsf{T} x) = (vv^\mathsf{T})x.$$

The expression $vv^\mathsf{T} \in \mathbb{R}^{m\times m}$ is called the *outer product* of the vector v with itself.

This formula completes the characterization of the *Householder reflection*.

Theorem 6.9 (Householder reflection). *Given a vector $x \in \mathbb{R}^m$, $x \neq 0$, the matrix*

$$\mathsf{H} = \mathsf{I} - 2\frac{vv^\mathsf{T}}{v^\mathsf{T} v}, \qquad v = x - \|x\|e_1,$$

is a symmetric orthogonal matrix in $\mathbb{R}^{m\times m}$ that maps the vector x onto the vector $\|x\|e_1$. Moreover,

$$\mathsf{H}^2 = \mathsf{I},$$

and all vectors orthogonal to v are invariant under multiplication by H.

Proof. By construction of the matrix H, we have

$$\mathsf{H}x = x - 2\frac{vv^\mathsf{T}}{v^\mathsf{T} v}x = \|x\|e_1,$$

and

$$\mathsf{H}^\mathsf{T} = \left(\mathsf{I} - 2\frac{vv^\mathsf{T}}{v^\mathsf{T} v}\right)^\mathsf{T} = \mathsf{I} - 2\frac{(vv^\mathsf{T})^\mathsf{T}}{v^\mathsf{T} v} = \mathsf{I} - 2\frac{vv^\mathsf{T}}{v^\mathsf{T} v} = \mathsf{H}.$$

The orthogonality of H follows from the observation that

$$\begin{aligned}
\mathsf{H}\mathsf{H}^\mathsf{T} = \mathsf{H}\mathsf{H} &= \left(\mathsf{I} - 2\frac{vv^\mathsf{T}}{v^\mathsf{T} v}\right)\left(\mathsf{I} - 2\frac{vv^\mathsf{T}}{v^\mathsf{T} v}\right) \\
&= \mathsf{I} - 2\frac{vv^\mathsf{T}}{v^\mathsf{T} v} - 2\frac{vv^\mathsf{T}}{v^\mathsf{T} v} + 4\frac{vv^\mathsf{T}}{v^\mathsf{T} v}\frac{vv^\mathsf{T}}{v^\mathsf{T} v} \\
&= \mathsf{I} - 4\frac{vv^\mathsf{T}}{v^\mathsf{T} v} + 4\frac{v(v^\mathsf{T} v)v^\mathsf{T}}{(v^\mathsf{T} v)^2} \\
&= \mathsf{I} - 4\frac{vv^\mathsf{T}}{v^\mathsf{T} v} + 4\frac{vv^\mathsf{T}}{v^\mathsf{T} v} = \mathsf{I},
\end{aligned}$$

which implies that

$$\mathsf{H}^{-1} = \mathsf{H}^{\mathsf{T}} = \mathsf{H}.$$

Finally, if $x \perp v$, then $\mathsf{H}x = x$, since $v^{\mathsf{T}}x = 0$. □

Returning to the orthogonal triangularization of the matrix A, we begin by computing the Householder reflector H_1 that sends the first column $a_1 \in \mathbb{R}^m$ of A to the vector with $\|a_1\|$ in the first position and zero in the entries below. The matrix $\mathsf{A}^{(1)} = \mathsf{H}_1\mathsf{A}$ is of the form

$$\mathsf{A}^{(1)} = \begin{bmatrix} \times & \times & \times & \times \\ 0 & \times & \times & \times \\ 0 & \times & \times & \times \\ 0 & \times & \times & \times \\ 0 & \times & \times & \times \end{bmatrix},$$

where, for the sake of visualization, we assumed that $A \in \mathbb{R}^{5\times 4}$, and the symbols "$\times$" indicate possible nonzero entries.

Next we consider the vector $x \in \mathbb{R}^{m-1}$ consisting of the last $m-1$ entries of the second column of $\mathsf{A}^{(1)}$, and we denote by $\widetilde{\mathsf{H}}_2 \in \mathbb{R}^{(m-1)\times(m-1)}$ the Householder reflector that maps x to the vector $\|x\|e_1$, where e_1 is the first canonical basis vector in $\mathbb{R}^{m-1}$. To get an orthogonal matrix that is compatible with $\mathsf{A}^{(1)}$ and preserves the zeros introduced by H_1 while introducing zeros in the second column below the diagonal entry, define

$$\mathsf{H}_2 = \begin{bmatrix} 1 & 0 \\ 0 & \widetilde{\mathsf{H}}_2 \end{bmatrix}.$$

It follows from the symmetry and orthogonality of $\widetilde{\mathsf{H}_2}$ that H_2 is symmetric and orthogonal, since

$$\mathsf{H}_2\mathsf{H}_2 = \begin{bmatrix} 1 & 0 \\ 0 & \widetilde{\mathsf{H}}_2 \end{bmatrix}\begin{bmatrix} 1 & 0 \\ 0 & \widetilde{\mathsf{H}}_2 \end{bmatrix} = \begin{bmatrix} 1 & 0 \\ 0 & \widetilde{\mathsf{H}}_2\widetilde{\mathsf{H}}_2 \end{bmatrix} = \begin{bmatrix} 1 & 0 \\ 0 & \mathsf{I}_{m-1} \end{bmatrix} = \mathsf{I}_m.$$

Moreover,

$$\mathsf{H}_2\mathsf{H}_1\mathsf{A} = \begin{bmatrix} \times & \times & \times & \times \\ 0 & \times & \times & \times \\ 0 & 0 & \times & \times \\ 0 & 0 & \times & \times \\ 0 & 0 & \times & \times \end{bmatrix}.$$

Continuing in this manner, we construct a sequence of orthogonal matrices $\mathsf{H}_1, \mathsf{H}_2, \ldots, \mathsf{H}_n$ such that

$$\mathsf{H}_1\mathsf{H}_2\cdots\mathsf{H}_n\mathsf{A} = \mathsf{R},$$

where R is an upper triangular matrix. We remark that if $n \geq m$ then $\mathsf{H}_m = \mathsf{H}_{m+1} = \cdots = \mathsf{H}_n = \mathsf{I}$. Since the product of orthogonal matrices is an orthogonal matrix, and the inverse of an orthogonal matrix is an orthogonal matrix given by its transpose,

$$\mathsf{A} = (\mathsf{H}_1\mathsf{H}_2\cdots\mathsf{H}_n)^{\mathsf{T}}\,\mathsf{R} = \mathsf{H}_n\mathsf{H}_{n-1}\cdots\mathsf{H}_1\mathsf{R},$$

and letting

$$\mathsf{Q} = \mathsf{H}_n\mathsf{H}_{n-1}\cdots\mathsf{H}_1,$$

we have the factorization of the matrix A,

$$\mathsf{A} = \mathsf{QR},$$

as the product of an orthogonal and an upper triangular matrix, thus proving Theorem 6.8.

Problems

1. Consider the mapping $F : \mathbb{R}^4 \to \mathbb{R}^2$, where

$$F \begin{bmatrix} v_1 \\ v_2 \\ v_3 \\ v_4 \end{bmatrix} = \begin{bmatrix} v_1 + v_2 \\ 3v_3 - 2v_4 \end{bmatrix}.$$

   Is F a linear mapping? Justify your answer.

2. Given the matrices $\mathsf{A} \in \mathbb{R}^{4\times 3}$, $\mathsf{B} \in \mathbb{R}^{3\times 4}$, and $\mathsf{C} \in \mathbb{R}^{5\times 2}$, which of the following products are defined, and for each product that is defined, how many rows and columns does the result have?

   (a) ABC
   (b) BAC
   (c) CAB
   (d) ACB
   (e) AB
   (f) BC
   (g) CB
   (h) AC
   (i) BA
   (j) CA

3. Given the matrices

$$\mathsf{A} = \begin{bmatrix} 3 \\ 0 \\ -1 \\ 5 \end{bmatrix}, \quad \mathsf{B} = \begin{bmatrix} 1 & 4 & 2 & 7 \end{bmatrix}, \quad \mathsf{C} = \begin{bmatrix} 2 & -1 & 0 & 4 \\ 0 & 1 & 0 & 1 \\ 2 & -3 & 1 & 0 \\ 0 & 0 & 2 & 3 \end{bmatrix},$$

   (a) Verify that $(\mathsf{AB})\,\mathsf{C} = \mathsf{A}\,(\mathsf{BC})$.
   (b) Using the definition of the matrix-matrix product, show that for the given matrices $(\mathsf{CA})^{\mathsf{T}} = \mathsf{A}^{\mathsf{T}}\mathsf{C}^{\mathsf{T}}$.
   (c) Compute BA and AB if they are both defined. Are they equal?

4. Using one of the definitions of the matrix-vector product, show that if for any $m \times n$ matrix $\mathsf{A} \in \mathbb{R}^{m\times n}$, there is a vector $x \neq 0$ such that $\mathsf{A}x = 0$, then the columns of A are not linearly independent.

5. Consider the linear mapping $F : \mathbb{R}^4 \to \mathbb{R}^3$,

$$F \begin{bmatrix} v_1 \\ v_2 \\ v_3 \\ v_4 \end{bmatrix} = \begin{bmatrix} v_1 + v_2 \\ v_3 - 4v_4 \\ v_1 - v_4 \end{bmatrix}.$$

   (a) Compute Fe_j, $j = 1, 2, 3, 4$, where e_1, e_2, e_3, e_4 are the canonical basis vectors of $\mathbb{R}^4$.

(b) Let A be the matrix whose columns are Fe_1, Fe_2, Fe_3, Fe_4, and let $v \in \mathbb{R}^4$ be the linear combination of e_1, e_2, e_3, e_4 with coefficients $2, -1, 7, -4$. Show that $Fv = \mathsf{A}c$, where $c \in \mathbb{R}^4$ is the vector whose entries are the components of v.

(c) Using the linearity of F, show that for any basis $\{w_1, w_2, w_4, w_4\}$ of $\mathbb{R}^4$, and for any vector $v \in \mathbb{R}^4$, $Fv = \mathsf{A}c$, where the jth column of A is Fw_j and c is the vector of the components of v in the basis $\{w_1, w_2, w_4, w_4\}$.

6. Prove that if the matrix $\mathsf{A} \in \mathbb{R}^{m \times m}$ is invertible and $\mathsf{A}x = \mathsf{A}y$, then $x = y$.

7. True or false? If true, prove it; if false, give a counterexample.

   (a) If A, B are $m \times m$ invertible matrices, AB and BA are also invertible. If the statement above is true, find the inverse of AB and BA in terms of those of A and B.

   (b) If A, B are invertible and of the same size, then their sum A + B is also invertible.

   (c) If A is invertible and $\mathsf{A}x = 0$, then $x = 0$.

   (d) If two $n \times n$ matrices A and B are symmetric, their products AB and BA are symmetric.

   (e) For any $m \times n$ matrix A, the products AA^T and $\mathsf{A}^\mathsf{T}\mathsf{A}$ are symmetric matrices.

8. Show that if Q is the product of orthogonal matrices, $\mathsf{Q} = \mathsf{Q}_1\mathsf{Q}_2 \ldots \mathsf{Q}_p$, then it is orthogonal, and find its inverse.

9. Using considerations about vector spaces and bounds on their dimensions,

   (a) Show that every 4×6 matrix has some nonzero vectors in its null space.

   (b) What can you say about the maximum dimension of the range of a 4×7 real matrix?

   (c) What can you say about the maximum dimension of the range of a 7×4 matrix?

10. Show that if $\mathsf{A} = \mathsf{QR}$, where $\mathsf{Q} \in \mathbb{R}^{m \times m}$ is an orthogonal matrix and $\mathsf{R} \in \mathbb{R}^{m \times n}$ is upper triangular, then the null space of A is equal to the null space of R.

11. Show that the range of an orthogonal matrix $\mathsf{Q} \in \mathbb{R}^{m \times m}$ is $\mathbb{R}^m$ and its null space is only the zero vector.

Chapter 7

Matrix Norms

Matrices can be thought of either for what they are, meaning that we regard them as a way of arranging their entries in a meaningful way, or for what they do, that is, focusing on how they change vectors they multiply. The way in which we measure a matrix should depend on which characteristics we are most interested in.

7.1 ▪ Matrices for what they do: Induced matrix norms

For the sake of definiteness, most of the discussion below will be restricted to real matrices, making the geometric interpretations more intuitive. However, the discussion extends with obvious modifications to complex matrices.

A way to assess the power of a matrix as an operator between vector spaces is to measure its stretching power. A vector $x \in \mathbb{R}^n$ upon multiplication by a matrix $\mathsf{A} \in \mathbb{R}^{m\times n}$ is mapped to a vector $y = \mathsf{A}x \in \mathbb{R}^m$. The first matrix norm that we introduce compares the norm of x to the norm of y.

Definition 7.1. *Let* $\mathbb{R}^n$ *and* $\mathbb{R}^m$ *be equipped with norms* $\|\cdot\|_{(n)} : \mathbb{R}^n \to \mathbb{R}_+$ *and* $\|\cdot\|_{(m)} : \mathbb{R}^m \to \mathbb{R}_+$, *respectively. The* induced norm *of the matrix* $\mathsf{A} \in \mathbb{R}^{m\times n}$ *is defined as*

$$\|\mathsf{A}\|_{(n,m)} = \sup_{x\neq 0} \frac{\|\mathsf{A}x\|_{(m)}}{\|x\|_{(n)}}.$$

In other words, the induced norm of a matrix measures the maximum dilation that a vector can undergo upon multiplication by the matrix. The induced norm is also referred to as operator norm.

Before presenting examples of induced matrix norms, a couple of comments are in order. First, let $x \neq 0$, and define

$$\widehat{x} = \frac{x}{\|x\|_{(n)}}, \quad \|\widehat{x}\|_{(n)} = \frac{1}{\|x\|_{(n)}}\|x\|_{(n)} = 1.$$

Observe that

$$\|\mathsf{A}\widehat{x}\|_{(m)} = \left\|\mathsf{A}\frac{x}{\|x\|_{(n)}}\right\|_{(m)} = \frac{\|\mathsf{A}x\|_{(m)}}{\|x\|_{(n)}};$$

that is, for each $x \neq 0$, there is an $\widehat{x}$ with $\|\widehat{x}\|_{(n)} = 1$ satisfying the above identity. Therefore we may define the induced matrix norm as

$$\|\mathsf{A}\|_{(n,m)} = \sup_{\|x\|_{(n)}=1} \|\mathsf{A}x\|_{(m)}.$$

To replace the supremum by a maximum, we need the following.

Theorem 7.2. *Let* A *be any* $m \times n$ *real matrix, and let* $\|\cdot\|_{(n)}$ *and* $\|\cdot\|_{(m)}$ *be norms in* $\mathbb{R}^n$ *and* $\mathbb{R}^m$*, respectively. The function*

$$g : \mathbb{R}^n \to \mathbb{R}^m, \quad g : x \mapsto \mathsf{A}x$$

is continuous in the sense that for every $\varepsilon > 0, x \in \mathbb{R}^n$*, there exists a* $\delta > 0$ *such that, for all* $z \in \mathbb{R}^n$, $\|z - x\|_{(n)} < \delta$ *implies* $\|\mathsf{A}x - \mathsf{A}z\|_{(m)} < \varepsilon$.

Proof. By the equivalence of norms in $\mathbb{R}^n$ and $\mathbb{R}^m$, we may restrict the proof to the 2-norm, or Euclidian norm, in both $\mathbb{R}^m$ and $\mathbb{R}^n$.

A vector $x \in \mathbb{R}^n$ can be expressed as the linear combination of the canonical basis vectors,

$$x = x_1 e_1 + \cdots + x_n e_n,$$

and, from the definition of a matrix-vector product,

$$\mathsf{A}x = x_1 \mathsf{A}e_1 + \cdots + x_n \mathsf{A}e_n.$$

The function g is bounded, because for any $x \in \mathbb{R}^n$,

$$\|g(x)\|_2 = \|\mathsf{A}x\|_2 \leq C\|x\|_2,$$

where

$$C = \sqrt{\|\mathsf{A}e_1\|_2^2 + \cdots + \|\mathsf{A}e_n\|_2^2}.$$

In fact, by the triangle inequality and the Cauchy–Schwarz inequality, we have

$$\begin{aligned} \|\mathsf{A}x\|_2 &\leq \|x_1 \mathsf{A}e_1\|_2 + \cdots + \|x_n \mathsf{A}e_n\|_2 \\ &= |x_1|\|\mathsf{A}e_1\|_2 + \cdots + |x_n|\|\mathsf{A}e_n\|_2 \\ &\leq C\|x\|_2. \end{aligned}$$

Given $x, y \in \mathbb{R}^n$, to guarantee that

$$\|\mathsf{A}x - \mathsf{A}y\|_2 < \varepsilon$$

for an arbitrary $\varepsilon > 0$, it suffices to require that

$$\|\mathsf{A}x - \mathsf{A}y\|_2 = \|\mathsf{A}(x - y)\|_2 \leq C\|x - y\|_2 < \varepsilon,$$

which holds if

$$\|x - y\|_2 < \delta = \frac{\varepsilon}{C},$$

completing the proof. □

It follows from the Extreme Value Theorem that a continuous function attains its maximum in a bounded and closed set of $\mathbb{R}^n$; therefore we may write

$$\|\mathsf{A}\|_{(n,m)} = \max_{\|x\|_{(n)}=1} \|\mathsf{A}x\|_{(m)}.$$

The verification that the induced matrix norm satisfies the norm axioms is left as an exercise (Problem 1).

A final comment concerning the choice of norms in the spaces $\mathbb{R}^n$ and $\mathbb{R}^m$ is in order. Usually they are chosen among the norms $\|\cdot\|_p$ with $p = 1, p = 2$, or $p = \infty$. When this is the case and p is the same for both spaces, it is customary to use the notation

$$\|\mathsf{A}\|_p = \max_{\|x\|_p=1} \|\mathsf{A}x\|_p, \quad p = 1, 2, \infty.$$

In the rest of this chapter we only consider induced norms of this type.

Before addressing how to evaluate the different induced matrix norms, we consider the special case of orthogonal matrices.

Theorem 7.3. *If* Q *is an orthogonal matrix, then* $\|\mathsf{Q}\|_2 = 1$.

Proof. By definition of orthogonal matrix,

$$\mathsf{Q}\mathsf{Q}^\mathsf{T} = \mathsf{Q}^\mathsf{T}\mathsf{Q} = \mathsf{I};$$

hence, for any vector x,

$$\|\mathsf{Q}x\|_2 = \|x\|_2,$$

and therefore,

$$\max_{x\neq 0} \frac{\|\mathsf{Q}x\|_2}{\|x\|_2} = \max_{x\neq 0} \frac{\|x\|_2}{\|x\|_2} = 1. \qquad \square$$

The same result holds for unitary matrices, with the transpose replaced by conjugate transpose. This is another way of saying that multiplication by orthogonal matrices does not change the length of vectors. An operation of this type is called an *isometry* of $\mathbb{R}^n$ or $\mathbb{C}^n$.

Unlike the induced 2-norm, the induced ∞- and 1-norms of matrices are very easy to evaluate.

Consider first the ∞-norm of a matrix. For simplicity, we start with a matrix consisting of a single row, that is,

$$\mathsf{A} = \begin{bmatrix} a_1 & a_2 & \cdots & a_n \end{bmatrix} \in \mathbb{R}^{1\times m}.$$

From the definition of induced norm, we need to maximize the value

$$|\mathsf{A}x| = \left| \sum_{j=1}^n x_j a_j \right|$$

over all vectors $x \in \mathbb{R}^n$ such that the entries satisfy $|x_j| \leq \max\{|x_j|\} = 1$. It is not hard to deduce that the maximum is attained if we choose x so that $x_j = 1$ when $a_j \geq 0$, and $x_j = -1$ if $a_j < 0$, leading to the conclusion that

$$\|\mathsf{A}\|_\infty = \sum_{j=1}^n |a_j|.$$

Therefore the infinity norm of a matrix with only one row is the absolute row sum.

By definition of infinity norm, if $\mathsf{A} \in \mathbb{R}^{m\times n}$,

$$\|\mathsf{A}\|_\infty = \max_{\|x\|_\infty=1} \{|(\mathsf{A}x)_1|, |(\mathsf{A}x)_2|, \ldots, |(\mathsf{A}x)_m|\},$$

where

$$\begin{bmatrix} (\mathsf{A}x)_1 \\ (\mathsf{A}x)_2 \\ \vdots \\ (\mathsf{A}x)_m \end{bmatrix} = \begin{bmatrix} a_{11}x_1 + a_{12}x_2 + \ldots + a_{1n}x_n \\ a_{21}x_1 + a_{22}x_2 + \ldots + a_{2n}x_n \\ \vdots \\ a_{m1}x_1 + a_{m2}x_2 + \ldots + a_{mn}x_n \end{bmatrix}.$$

Applying the reasoning above to each row, it turns out that the vector x that maximizes the maximum of the $|(\mathsf{A}x)_j|$ is the one that produces the largest absolute row sum. This x is the vector with entries ± 1 corresponding to the signs of the entries in the row with the largest absolute row sum. Therefore we conclude that

$$\|\mathsf{A}\|_\infty = \max_{1\le i\le m} \sum_{j=1}^{n} |a_{ij}|.$$

Observe that in case the matrix A has only one column, the maximum row sum rule picks the largest absolute entry, hence coinciding with the infinity norm of the column as a vector.

For the computation of $\|\mathsf{A}\|_1$, we begin with the observation that if $\|x\|_1 = 1$, then, by definition,

$$|x_1| + |x_2| + \cdots + |x_n| = 1,$$

and

$$\begin{aligned} \|\mathsf{A}\|_1 &= \max_{\|x\|_1=1} \{\|\mathsf{A}x\|_1\} \\ &= \max_{\|x\|_1=1} \left(|(\mathsf{A}x)_1| + \cdots + |(\mathsf{A}x)_m|\right). \end{aligned}$$

Recalling the row-wise interpretation of matrix-vector product, the expression to maximize is

$$\begin{aligned} &|(\mathsf{A}x)_1| + \cdots + |(\mathsf{A}x)_m| \\ &\quad = |a_{11}x_1 + a_{12}x_2 + \cdots + a_{1n}x_n| + \cdots + |a_{m1}x_1 + a_{m2}x_2 + \cdots + a_{mn}x_n| \\ &\quad \le \big(|x_1||a_{11}| + |x_2||a_{12}| + \cdots + |x_n||a_{1n}|\big) + \cdots \\ &\qquad + \big(|x_1||a_{m1}| + |x_2||a_{m2}| + \cdots + |x_n||a_{mn}|\big) \\ &\quad = |x_1|\big(|a_{11}| + |a_{21}| + \cdots + |a_{m1}|\big) + \cdots + |x_n|\big(|a_{1n}| + |a_{2n}| + \cdots + |a_{mn}|\big). \end{aligned}$$

If c_j denotes the jth absolute column sum,

$$c_j = |a_{1j}| + |a_{2j}| + \cdots + |a_{mj}|, \quad 1 \le j \le n,$$

then

$$|(\mathsf{A}x)_1| + \cdots + |(\mathsf{A}x)_m| \le |x_1|c_1 + |x_2|c_2 + \cdots + |x_n|c_n \le \max_{1\le j\le n} c_j,$$

that is,

$$\|\mathsf{A}\|_1 \le \max_{1\le j\le n} c_j.$$

Let j^* be the index that corresponds to the maximum column sum, that is,

$$\max_{1\le j\le n} c_j = c_{j^*}.$$

If we choose $x = e_{j^*}$, the canonical coordinate vector, we observe that

$$\|\mathsf{A}x\|_1 = |(\mathsf{A}x)_1| + \cdots + |(\mathsf{A}x)_m| = |a_{1j^*}| + |a_{2j^*}| + \cdots + |a_{mj^*}| = c_{j^*},$$

demonstrating that the upper bound is attained. We therefore conclude that

$$\|\mathsf{A}\|_1 = \max_{1\le j\le n} \sum_{k=1}^{m} |a_{kj}|.$$

Unfortunately there is no formula for computing the 2-norm of a generic matrix, although it is possible to infer it for some classes of matrices.

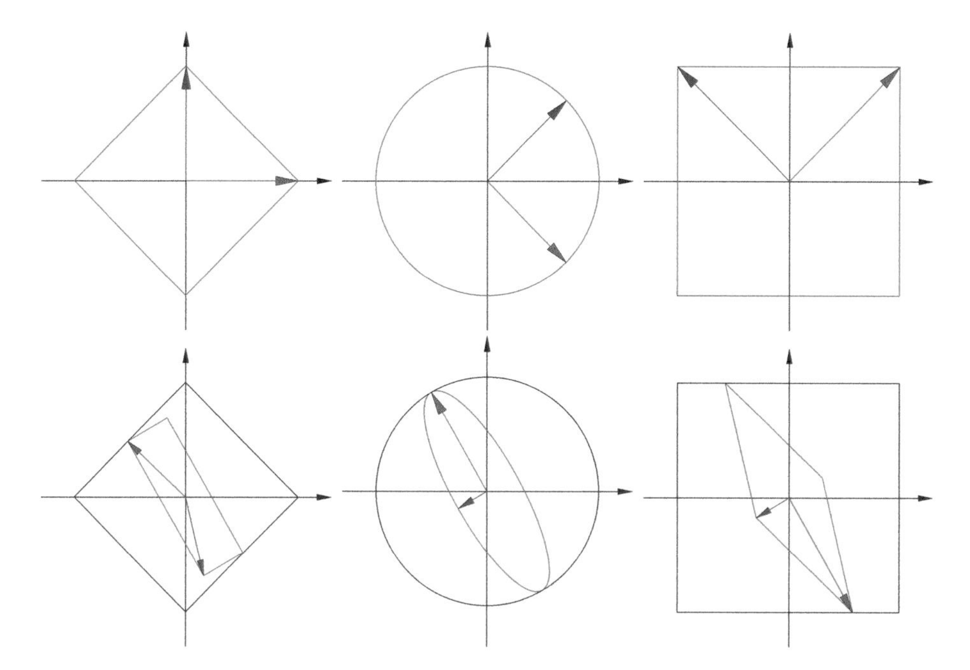

Figure 7.1. *The upper row shows the unit spheres in* 1*-norm,* 2*-norm, and* ∞*-norm. In the lower row, the images of the unit spheres under an action of the same* 2×2 *matrix are plotted in red. The maximum norm over the image of the unit sphere is found by drawing the smallest sphere that contains the image of the unit sphere, and the vector corresponding to the maximum is indicated by a red arrow. Observe that to find the maximum with respect to* 1*-norm of* ∞*-norm, in* $\mathbb{R}^2$ *only two candidate vectors need to be tested, while in the case of* 2*-norm, the rotational symmetry of the sphere makes the question more complicated, and no simple algorithm for computing the corresponding matrix norm can be given.*

It is insightful to think about matrix norms in geometric terms. As pointed out before, the unit spheres with respect to the 1- and ∞-norms are polytopes, and due to the linearity of the mapping effectuated by the matrix multiplication, the images of the spheres are also polytopes. To find the vector in the image set that maximizes the norm, it suffices to consider only the vertices of the polytopes, and by symmetry, it suffices to choose the largest, as illustrated in Figure 7.1.

7.2 ▪ Matrices for what they are: The Frobenius norm

The arrangement of the entries of a matrix into rows and columns is often a convenient way of keeping track of quantities. In other words, a matrix may be used only as an organizer of values such as multivariable data, and its action on vectors is less important. As an example, we may consider an image as a matrix, where the entry a_{ij} is the grayscale value of the pixel in the position (i, j). This interpretation of a matrix for *what it is* calls for a norm that measures the entries of the matrix itself. The most popular norm of this type is the *Frobenius norm*, which resembles the 2-norm for vectors.

Definition 7.4. *The Frobenius norm of the* $m \times n$ *real or complex matrix* A *is defined as*

$$\|\mathsf{A}\|_{\mathrm{F}} = \left(\sum_{i=1}^{m} \sum_{j=1}^{n} |a_{ij}|^2 \right)^{1/2} .$$

The Frobenius norm can be expressed in different ways that may be very convenient in certain cases. If we partition a matrix into its column vectors,

$$\mathsf{A} = \begin{bmatrix} \vdots & \vdots & & \vdots \\ a_1 & a_2 & \cdots & a_n \\ \vdots & \vdots & & \vdots \end{bmatrix}, \qquad a_j \in \mathbb{R}^m,$$

the relation between the Frobenius norm of the matrix and the 2-norm of its columns,

$$\|\mathsf{A}\|_{\mathrm{F}}^2 = \sum_{j=1}^{n} \|a_j\|_2^2,$$

is rather obvious. Similarly, since $\|\mathsf{A}\|_{\mathrm{F}} = \|\mathsf{A}^{\mathsf{T}}\|_{\mathrm{F}}$, we may express the square of the Frobenius norm as the sum of the squared 2-norms of the row vectors of A.

A different measure for the size of the matrix is given by the trace. The *trace* of an $n \times n$ matrix is the sum of the elements on its main diagonal,

$$\mathrm{trace}(\mathsf{A}) = \sum_{j=1}^{n} a_{jj}.$$

The following theorem establishes a connection between the trace and the Frobenius norm.

Theorem 7.5. *Let* A *and* B *be* $n \times n$ *matrices. Then*

$$\mathrm{trace}(\mathsf{AB}) = \mathrm{trace}(\mathsf{BA}),$$

and for $\mathsf{A} \in \mathbb{R}^{m \times n}$,

$$\|\mathsf{A}\|_{\mathrm{F}}^2 = \mathrm{trace}\left(\mathsf{A}^{\mathsf{T}}\mathsf{A}\right) = \mathrm{trace}\left(\mathsf{A}\mathsf{A}^{\mathsf{T}}\right).$$

Proof. For square matrices A and B,

$$(\mathsf{AB})_{ij} = \sum_{k=1}^{n} a_{ik}b_{kj}, \quad (\mathsf{BA})_{ij} = \sum_{\ell=1}^{n} b_{i\ell}a_{\ell j};$$

therefore

$$\mathrm{trace}(\mathsf{AB}) = \sum_{\ell=1}^{n}\left(\sum_{k=1}^{n} a_{\ell k}b_{k\ell}\right) = \sum_{k=1}^{n}\left(\sum_{\ell=1}^{n} b_{k\ell}a_{\ell k}\right) = \mathrm{trace}(\mathsf{BA}).$$

Furthermore, for $\mathsf{A} \in \mathbb{R}^{m \times n}$,

$$\left(\mathsf{A}^{\mathsf{T}}\mathsf{A}\right)_{ij} = \sum_{k=1}^{m} a_{ki}a_{kj}, \quad \left(\mathsf{A}\mathsf{A}^{\mathsf{T}}\right)_{ij} = \sum_{k=1}^{n} a_{ik}a_{jk};$$

therefore

$$\begin{aligned} \mathrm{trace}(\mathsf{A}^{\mathsf{T}}\mathsf{A}) &= \sum_{j=1}^{n}\left(\sum_{k=1}^{m} a_{kj}a_{kj}\right) = \|\mathsf{A}\|_{\mathrm{F}}^2 \\ &= \sum_{k=1}^{m}\left(\sum_{j=1}^{m} a_{kj}a_{kj}\right) = \mathrm{trace}(\mathsf{A}\mathsf{A}^{\mathsf{T}}), \end{aligned}$$

as claimed. □

Problems

1. Prove that the induced matrix norm satisfies the properties required from a norm in a vector space.

2. There are two useful properties of induced matrix norms: if $\mathsf{A} \in \mathbb{R}^{m\times n}$ and $\mathsf{B} \in \mathbb{R}^{n\times k}$ are matrices and $x \in \mathbb{R}^n$ is a vector, then

$$\|\mathsf{A}x\| \leq \|\mathsf{A}\|\|x\|$$

and

$$\|\mathsf{AB}\| \leq \|\mathsf{A}\|\|\mathsf{B}\|.$$

Show these inequalities using the definition of induced matrix norms. You may assume that the norms are the p-norms with $p = 1$, $p = 2$, or $p = \infty$.

3. Compute the ∞-, 1-, and Frobenius norm of the matrix

$$\begin{bmatrix} 2 & 4 & -11 & 4 & -9 \\ 4 & -1 & 6 & 2 & -1 \\ 1 & 5 & 6 & -7 & -8 \\ -20 & 20 & 2 & -2 & 0 \end{bmatrix}.$$

4. If $\mathsf{A} \in \mathbb{R}^{m\times n}$, and $\mathsf{P} \in \mathbb{R}^{m\times m}$, $\mathsf{Q} \in \mathbb{R}^{n\times n}$ are orthogonal matrices, prove that

$$\|\mathsf{PA}\|_2 = \|\mathsf{A}\|_2, \qquad \|\mathsf{AQ}\|_2 = \|\mathsf{A}\|_2.$$

5. For matrices of size $m \times n$, does the assignment

$$\max(\mathsf{A}) = \max\{a_{ij}\}$$

define a norm in the vector space of $m \times n$ matrices? What about the assignment

$$\max(|\mathsf{A}|) = \max\{|a_{ij}|\}?$$

Justify your answer.

6. Given the matrix

$$\mathsf{A} = \begin{bmatrix} 2 & 0 & 3 & -6 \\ 4 & 3 & 0 & 0 \\ 4 & 1 & -1 & -1 \\ 0 & 3 & 0 & 2 \end{bmatrix},$$

   (a) If you know that $\|\mathsf{A}\|_2 = 6.9775$, and $\mathsf{A} = \mathsf{QR}$ is the QR factorization of A, find $\|\mathsf{R}\|_2$. Justify your answer.

   (b) Compute $\|\mathsf{A}\|_1$ and $\|\mathsf{A}^\mathsf{T}\|_1$.

   (c) Compute $\|\mathsf{A}\|_\infty$ and $\|\mathsf{A}^\mathsf{T}\|_\infty$.

   (d) Compute $\|\mathsf{A}\|_\mathrm{F}$ and $\|\mathsf{A}^\mathsf{T}\|_\mathrm{F}$.

   (e) Compute $\mathsf{A}^\mathsf{T}\mathsf{A}$ and $\mathsf{A}\mathsf{A}^\mathsf{T}$, and verify that

$$\operatorname{trace}\left(\mathsf{A}^\mathsf{T}\mathsf{A}\right) = \operatorname{trace}\left(\mathsf{A}\mathsf{A}^\mathsf{T}\right) = \|\mathsf{A}\|_\mathrm{F}^2.$$

   (f) If $\mathsf{A} = \mathsf{QR}$ is the QR-factorization of the $n \times n$ matrix A, show that A is invertible if and only if R is invertible, and write the inverse of A in terms of its QR factors.

7. True or false? Justify your answer.

   (a) If Q is an orthogonal matrix, then

$$\|\mathsf{QA}\|_{\mathrm{F}} = \|\mathsf{A}\|_{\mathrm{F}}.$$

   (b) If Q is an orthogonal matrix, then

$$\|\mathsf{AQ}\|_{\mathrm{F}} = \|\mathsf{A}\|_{\mathrm{F}}.$$

Chapter 8

The Singular Value Decomposition

The singular value decomposition (SVD) is, arguably, the most powerful matrix factorization in linear algebra. Unlike other factorizations that may be applicable only to restricted classes of matrices, the SVD, the most democratic of factorizations, exists regardless of the dimensions of the matrix.

8.1 ▪ Basis completion

We start with a technical result showing that a few orthonormal vectors in a vector space V equipped with an inner product can always be completed into an orthonormal basis of V.

Theorem 8.1. *Given k orthonormal vectors $q_1, \dots, q_k$ in a vector space V of dimension n with inner product $\langle \cdot , \cdot \rangle$, with $k \le n$, it is always possible to find vectors $q_{k+1}, \dots, q_n$ such that $q_1, \dots, q_n$ is an orthonormal basis of V.*

Proof. The proof follows the lines of the Gram–Schmidt orthogonalization.

Every basis of the vector space V contains n vectors; therefore, if $k = n$, there is nothing to prove. If $k < n$, then the vectors $\{q_1, \dots, q_k\}$ are not a spanning set, and consequently there must be a vector $v \neq 0$, $v \notin \mathrm{span}\{q_1, q_2, \dots, q_k\}$, such that

$$z = v - \langle v, q_1 \rangle q_1 - \cdots - \langle v, q_k \rangle q_k \neq 0.$$

The vector $q_{k+1} = \frac{z}{\|z\|}$ has unit length, and for each j, $1 \le j \le k$, by the orthonormality of the vectors we have

$$\begin{aligned} \langle z, q_j \rangle &= \langle v, q_j \rangle - \langle v, q_1 \rangle \langle q_1, q_j \rangle - \cdots - \langle v, q_k \rangle \langle q_k, q_j \rangle \\ &= \langle v, q_j \rangle - \langle v, q_j \rangle \langle q_j, q_j \rangle \\ &= 0. \end{aligned}$$

Therefore we conclude that $q_1, \dots, q_k, q_{k+1}$ is an orthonormal set. If $k + 1 = n$, the claim is proved; if $k+1 < n$, we repeat the argument until the desired orthonormal basis is found. □

8.2 ▪ Existence of the SVD

Before proving the existence of the SVD, we extend the definition of diagonal matrices. A matrix $\Sigma \in \mathbb{R}^{m \times n}$ is diagonal if the entries s_{ij} vanish for $i \neq j$. If Σ is an $m \times n$ non–square diagonal matrix, it contains at least $(n-m)$ zero columns if $n > m$, and at least $(m-n)$ zero rows if $m > n$, as indicated by the matrix schematics

$$\Sigma = \left[\begin{array}{ccc|c} \times & & & \\ & \ddots & & \mathsf{O}_{m\times(n-m)} \\ & & \times & \end{array}\right], \text{ or } \Sigma = \left[\begin{array}{ccc} \times & & \\ & \ddots & \\ & & \times \\ \hline & \mathsf{O}_{(m-n)\times n} & \end{array}\right],$$

where $\mathsf{O}_{k\times\ell}$ denotes a zero matrix of the indicated size $k \times \ell$. In general, the maximum number of nonvanishing diagonal entries is $\min\{m,n\}$.

The following theorem states one of the most important results in linear algebra.

Theorem 8.2 (Existence of the singular value decomposition). *For any matrix* $\mathsf{A} \in \mathbb{R}^{m\times n}$, *there exist an* $m\times m$ *orthogonal matrix* U, *an* $n\times n$ *orthogonal matrix* V, *and an* $m\times n$ *diagonal matrix* Σ *with nonnegative diagonal entries* $\sigma_1 \geq \sigma_2 \geq \cdots \geq \sigma_{\min\{m,n\}} \geq 0$ *such that*

$$\mathsf{A} = \mathsf{U}\Sigma\mathsf{V}^{\mathsf{T}}. \tag{8.1}$$

Proof. We start by denoting the induced 2-norm of the matrix A by σ_1. If $\sigma_1 = 0$, then by the first property of norms the matrix A is the zero matrix, and the statement holds with U, V identity matrices of the appropriate sizes, and $\Sigma = \mathsf{A}$.

If $\sigma_1 \neq 0$, by the definition of induced 2-norm there exists a vector $v_1 \in \mathbb{R}^n$ with $\|v_1\|_2 = 1$ such that $\mathsf{A}v_1 = y$, with $\|y\|_2 = \sigma_1 > 0$. It follows from the properties of norms that $y \neq 0$ is nonzero and the vector $u_1 = \sigma_1^{-1} y \in \mathbb{R}^m$ has unit 2-norm. Theorem 8.1 guarantees that we can select $m-1$ additional vectors $\widetilde{u}_j$, $2 \leq j \leq m$, to form an orthonormal basis for $\mathbb{R}^m$. Denote by U_1 the orthogonal matrix whose columns are the vectors in the orthonormal basis just constructed,

$$\mathsf{U}_1 = \begin{bmatrix} \vdots & \vdots & & \vdots \\ u_1 & \widetilde{u}_2 & \cdots & \widetilde{u}_m \\ \vdots & \vdots & & \vdots \end{bmatrix} \in \mathbb{R}^{m\times m}.$$

Similarly, starting with the vector $v_1 \in \mathbb{R}^n$ of unit norm, we can construct an orthonormal basis for $\mathbb{R}^n$. The matrix V_1, whose columns are the orthonormal basis vectors,

$$\mathsf{V}_1 = \begin{bmatrix} \vdots & \vdots & & \vdots \\ v_1 & \widetilde{v}_2 & \cdots & \widetilde{v}_n \\ \vdots & \vdots & & \vdots \end{bmatrix} \in \mathbb{R}^{n\times n},$$

is also an orthogonal matrix.

By construction,

$$\mathsf{A}v_1 = y = \sigma_1 u_1;$$

therefore

$$\mathsf{A}\mathsf{V}_1 = \begin{bmatrix} \vdots & \vdots & & \vdots \\ \mathsf{A}v_1 & \mathsf{A}\widetilde{v}_2 & \cdots & \mathsf{A}\widetilde{v}_n \\ \vdots & \vdots & & \vdots \end{bmatrix} = \begin{bmatrix} \vdots & \vdots & & \vdots \\ \sigma_1 u_1 & \mathsf{A}\widetilde{v}_2 & \cdots & \mathsf{A}\widetilde{v}_n \\ \vdots & \vdots & & \vdots \end{bmatrix}.$$

Multiplying the matrix AV_1 by U_1^T from the left, and using the orthogonality of u_1 to all the subsequent columns of U_1, we observe that

$$\mathsf{U}_1^\mathsf{T}\mathsf{A}v_1 = \mathsf{U}_1^\mathsf{T}(\sigma u_1) = \begin{bmatrix} \sigma_1 u_1^\mathsf{T} u_1 \\ \sigma_1 \widetilde{u}_2^\mathsf{T} u_1 \\ \vdots \\ \sigma_1 \widetilde{u}_m^\mathsf{T} u_1 \end{bmatrix} = \begin{bmatrix} \sigma_1 \\ 0 \\ \vdots \\ 0 \end{bmatrix}.$$

Therefore,

$$\mathsf{U}_1^\mathsf{T}\mathsf{AV}_1 = \begin{bmatrix} \sigma_1 & \vdots & & \vdots \\ 0 & & & \\ \vdots & \mathsf{U}_1^\mathsf{T}\mathsf{A}\widetilde{v}_2 & \dots & \mathsf{U}_1^\mathsf{T}\mathsf{A}\widetilde{v}_n \\ 0 & \vdots & & \vdots \end{bmatrix} = \left[\begin{array}{c|c} \sigma_1 & w^\mathsf{T} \\ \hline 0 & \\ \vdots & \mathsf{B} \\ 0 & \end{array}\right],$$

where $w \in \mathbb{R}^{n-1}$ and $\mathsf{B} \in \mathbb{R}^{(m-1)\times(n-1)}$. Next we show that $w = 0$. To that end, observe that because U_1 and V_1 are orthogonal matrices,

$$\|\mathsf{U}_1^\mathsf{T}\mathsf{AV}_1\|_2 = \|\mathsf{A}\|_2 = \sigma_1.$$

Therefore, denoting

$$\mathsf{A}_1 = \mathsf{U}_1^\mathsf{T}\mathsf{AV}_1,$$

we have that, for any $p \in \mathbb{R}^n$, $p \neq 0$,

$$\sigma_1 = \|\mathsf{A}_1\|_2 = \max_{v\neq 0} \frac{\|\mathsf{A}_1 v\|_2}{\|v\|_2} \geq \frac{\|\mathsf{A}_1 p\|_2}{\|p\|_2}.$$

Let

$$p = \left[\begin{array}{c} \sigma_1 \\ \hline \vdots \\ w \\ \vdots \end{array}\right] \in \mathbb{R}^n,$$

and observe that

$$\|p\|_2 = \sqrt{p^\mathsf{T}p} = \sqrt{\sigma_1^2 + w^\mathsf{T}w}.$$

Then

$$\mathsf{A}_1 p = \left[\begin{array}{c|c} \sigma_1 & w^\mathsf{T} \\ \hline 0 & \\ \vdots & \mathsf{B} \\ 0 & \end{array}\right] \left[\begin{array}{c} \sigma_1 \\ \hline \vdots \\ w \\ \vdots \end{array}\right] = \left[\begin{array}{c} \sigma_1^2 + w^\mathsf{T}w \\ \hline \vdots \\ \mathsf{B}w \\ \vdots \end{array}\right],$$

and from

$$w^\mathsf{T}\mathsf{B}^\mathsf{T}\mathsf{B}w = \|\mathsf{B}w\|_2^2 \geq 0,$$

it follows that

$$\begin{aligned} \|\mathsf{A}_1 p\|_2 &= \sqrt{\left(\sigma_1^2 + w^\mathsf{T}w\right)^2 + w^\mathsf{T}\mathsf{B}^\mathsf{T}\mathsf{B}w} \\ &\geq \sigma_1^2 + w^\mathsf{T}w. \end{aligned}$$

In light of the observation above, if $w \neq 0$, then

$$\frac{\|\mathsf{A}_1 p\|_2}{\|p\|_2} \geq \frac{\sigma_1^2 + w^\mathsf{T} w}{\sqrt{\sigma_1^2 + w^\mathsf{T} w}} = \sqrt{\sigma_1^2 + w^\mathsf{T} w} > \sigma_1,$$

contradicting the fact that

$$\frac{\|\mathsf{A}_1 p\|_2}{\|p\|_2} \leq \sigma_1.$$

Therefore we conclude that $w = 0$, and

$$\mathsf{A}_1 = \left[\begin{array}{c|ccc} \sigma_1 & 0 & \cdots & 0 \\ \hline 0 & & & \\ \vdots & & \mathsf{B} & \\ 0 & & & \end{array}\right].$$

We can now apply the same procedure to the matrix B and define the orthogonal matrices

$$\mathsf{U}_2 = \left[\begin{array}{c|ccc} \mathsf{I}_1 & 0 & \cdots & 0 \\ \hline 0 & & & \\ \vdots & & \widetilde{\mathsf{U}}_2 & \\ 0 & & & \end{array}\right] \in \mathbb{R}^{m \times m}, \quad \mathsf{V}_2 = \left[\begin{array}{c|ccc} \mathsf{I}_1 & 0 & \cdots & 0 \\ \hline 0 & & & \\ \vdots & & \widetilde{\mathsf{V}}_2 & \\ 0 & & & \end{array}\right] \in \mathbb{R}^{n \times n},$$

where $\widetilde{\mathsf{U}}_2 \in \mathbb{R}^{(m-1)\times(m-1)}$ and $\widetilde{\mathsf{V}}_2 \in \mathbb{R}^{(n-1)\times(n-1)}$ are orthogonal matrices such that

$$\widetilde{\mathsf{U}}_2^\mathsf{T} \mathsf{B} \widetilde{\mathsf{V}}_2 = \mathsf{B}_1 = \left[\begin{array}{c|ccc} \sigma_2 & 0 & \cdots & 0 \\ \hline 0 & & & \\ \vdots & & \mathsf{C} & \\ 0 & & & \end{array}\right],$$

with

$$\sigma_2 = \|\mathsf{B}_1\|_2.$$

It is not hard to see that we must have $\sigma_2 \leq \sigma_1$. Indeed, if $\widetilde{y} \in \mathbb{R}^{n-1}$ is a vector with $\|\widetilde{y}\|_2 = 1$ and $\|\mathsf{B}\widetilde{y}\|_2 = \sigma_2$, and

$$y = \left[\begin{array}{c} 0 \\ \hline \vdots \\ \widetilde{y} \\ \vdots \end{array}\right],$$

then $\|y\|_2 = 1$; thus, since $\|\mathsf{A}_1\|_2 = \sigma_1$,

$$\sigma_1 \geq \|\mathsf{A}_1 y\|_2 = \|\mathsf{B}\widetilde{y}\|_2 = \sigma_2.$$

The process continues in a similar fashion until the matrix has been reduced to a diagonal form. If $m \geq n$, this requires at most $(n-1)$ steps and defines two sequences of orthogonal matrices $\mathsf{V}_1, \ldots, \mathsf{V}_{n-1} \in \mathbb{R}^{n \times n}$ and $\mathsf{U}_1, \ldots, \mathsf{U}_{n-1} \in \mathbb{R}^{m \times m}$ such that

$$\mathsf{U}_{n-1}^\mathsf{T} \mathsf{U}_{n-2}^\mathsf{T} \cdots \mathsf{U}_1^\mathsf{T} \mathsf{A} \mathsf{V}_1 \mathsf{V}_2 \cdots \mathsf{V}_{n-1} = \Sigma,$$

where

$$\Sigma = \left[\begin{array}{ccc} \sigma_1 & & \\ & \ddots & \\ & & \sigma_n \\ \hline & \mathsf{O}_{(m-n)\times n} & \end{array}\right]$$

is an $m \times n$ diagonal matrix and

$$\sigma_1 \geq \sigma_2 \geq \cdots \geq \sigma_n \geq 0.$$

Letting

$$\mathsf{U}^\mathsf{T} = \mathsf{U}_{n-1}^\mathsf{T}\mathsf{U}_{n-2}^\mathsf{T}\cdots\mathsf{U}_1^\mathsf{T} = (\mathsf{U}_1\mathsf{U}_2\cdots\mathsf{U}_{n-1})^\mathsf{T}$$

and

$$\mathsf{V} = \mathsf{V}_1\mathsf{V}_2\cdots\mathsf{V}_{n-1}$$

and recalling that the product of orthogonal matrices is an orthogonal matrix, we have that

$$\mathsf{A} = \mathsf{U}\Sigma\mathsf{V}^\mathsf{T},$$

hence completing the proof. The proof for case $m \leq n$ is similar. □

The factorization (8.1) is called the *singular value decomposition* of A. The diagonal entries of Σ are the *singular values* of A, and the columns of the matrices U and V are called the *left* and *right singular vectors*, respectively.

The proof above shows that as long as the singular values are strictly decreasing, the SVD of a matrix $\mathsf{A} \in \mathbb{R}^{m\times n}$ is unique up to a sign scaling of the singular vector pairs (u_j, v_j) constituting the columns of U and V. If some of the singular values coincide, more ambiguity ensues, as can be seen, e.g., by considering the decomposition of the identity matrix as a product,

$$\mathsf{I} = \mathsf{U}\mathsf{I}\mathsf{U}^\mathsf{T},$$

where U is any orthogonal matrix. This is the SVD of I, having all singular values equal to one.

The results of Theorem 8.2 extend in a natural way to complex matrices, with the matrices U and V unitary instead of orthogonal, and transposition replaced by conjugate transposition. Similarly, as long as the singular values do not coincide, the uniqueness of the SVD is only up to a scaling of the pairs of singular vectors (u_j, v_j) by a factor of modulus one.

The singular values and singular vectors of a matrix A have the important property that

$$\mathsf{A}v_j = \sigma_j u_j, \quad 1 \leq j \leq \min\{n, m\}.$$

Moreover, if $m < n$, that is, if there are more right singular vectors v_j than left singular vectors u_j, we have

$$\mathsf{A}v_j = 0 \text{ for } m < j \leq n.$$

The multiplication of a vector x by a matrix A is not, in general, immediately interpretable, unless the matrix is diagonal. In the following we show how the SVD makes it possible to think *diagonally* about any matrix A. Let $\mathsf{A} \in \mathbb{R}^{m\times n}$. Replace A by its SVD to express the product $\mathsf{A}x$ in the form

$$\mathsf{U}\Sigma\mathsf{V}^\mathsf{T}x = y. \tag{8.2}$$

Multiplying both sides of (8.2) from the left by U^T and using the orthogonality of U, we get

$$\Sigma\mathsf{V}^\mathsf{T}x = \mathsf{U}^\mathsf{T}y.$$

Multiplication of vectors in $\mathbb{R}^n$ by an orthogonal matrix V^T does not change their length or the angle between any pair of vectors. This implies that the matrix V^T defines an isometric change of coordinates in the vector space $\mathbb{R}^n$, and the matrix U^T defines an isometric change of coordinates in $\mathbb{R}^m$. Denoting the new coordinates by $y' = \mathsf{U}^\mathsf{T} y$ and $x' = \mathsf{V}^\mathsf{T} x$, and observing that

$$\mathsf{U}^\mathsf{T} y = y' \Leftrightarrow \mathsf{U} y' = y, \quad \mathsf{V}^\mathsf{T} x = x' \Leftrightarrow \mathsf{V} x' = x,$$

we see that multiplication of a vector x by the matrix A is the same as multiplication of the vector x' by the diagonal matrix Σ ,

$$\mathsf{A} x = y \Leftrightarrow \Sigma x' = y'.$$

The diagonal representation of the action of A shows that when a vector x is multiplied by A, the kth component of x', $k \leq \min\{m, n\}$, is scaled by σ_k, the kth singular value of A. Thus, for $m \leq n$,

$$\Sigma x' = \Sigma \begin{bmatrix} x_1' \\ \vdots \\ x_n' \end{bmatrix} = \begin{bmatrix} \sigma_1 x_1' \\ \vdots \\ \sigma_m x_m' \end{bmatrix},$$

and for $m > n$,

$$\Sigma x' = \Sigma \begin{bmatrix} x_1' \\ \vdots \\ x_n' \end{bmatrix} = \begin{bmatrix} \sigma_1 x_1' \\ \vdots \\ \sigma_n x_n' \\ 0 \\ \vdots \\ 0 \end{bmatrix}.$$

Example 10: To explain the significance of the SVD in geometric terms, consider a matrix $\mathsf{A} \in \mathbb{R}^{2\times 2}$ given in terms of its SVD,

$$\mathsf{A} = \mathsf{U}\Sigma\mathsf{V}^\mathsf{T},$$

where

$$\mathsf{V} = \begin{bmatrix} -1/\sqrt{2} & -1/\sqrt{2} \\ 1/\sqrt{2} & -1/\sqrt{2} \end{bmatrix} = \begin{bmatrix} \cos\theta & -\sin\theta \\ \sin\theta & \cos\theta \end{bmatrix}, \quad \theta = \frac{3\pi}{4},$$

$$\Sigma = \begin{bmatrix} 6/5 & 0 \\ 0 & 1/2 \end{bmatrix} = \begin{bmatrix} \sigma_1 & 0 \\ 0 & \sigma_2 \end{bmatrix},$$

and

$$\mathsf{U} = \begin{bmatrix} 1/2 & -\sqrt{3}/2 \\ \sqrt{3}/2 & 1/2 \end{bmatrix} = \begin{bmatrix} \cos\varphi & -\sin\varphi \\ \sin\varphi & \cos\varphi \end{bmatrix}, \quad \varphi = \frac{\pi}{3}.$$

Let's begin by considering the matrix V. Observe that

$$v_1 = \mathsf{V} e_1 = \begin{bmatrix} \cos\theta \\ \sin\theta \end{bmatrix}, \quad v_2 = \mathsf{V} e_2 = \begin{bmatrix} -\sin\theta \\ \cos\theta \end{bmatrix};$$

that is, V represents the counterclockwise rotation of the plane around the origin by the angle θ. Since $\mathsf{V}^\mathsf{T} = \mathsf{V}^{-1}$, we have

$$\mathsf{V}^\mathsf{T} v_1 = e_1, \quad \mathsf{V}^\mathsf{T} v_2 = e_2,$$

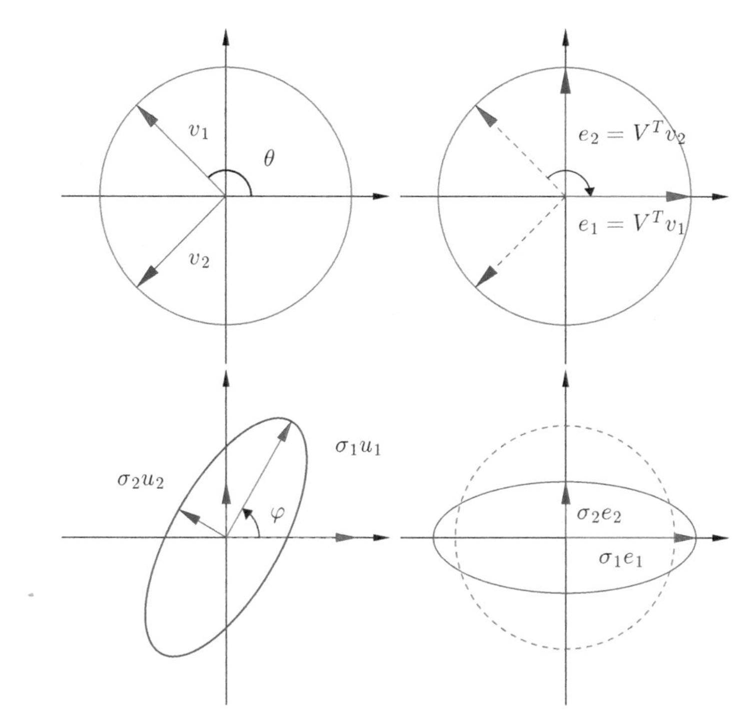

Figure 8.1. *The four panels, in clockwise order, provide a graphical illustration of the action of matrix multiplication using the SVD. Upper left: The column vectors of the matrix* V. *Upper right: Multiplication by the matrix* $\mathsf{V}^\mathsf{T} = \mathsf{V}^{-1}$ *is a rotation that aligns the column vectors* v_1 *and* v_2 *along the coordinate axes. Lower right: Scaling of the canonical basis vectors by the singular values. Lower left: Rotation of the scaled canonical basis vectors by* U *aligns them with the columns* u_1 *and* u_2 *of the matrix* U.

where multiplication by V^T is equivalent to rotating a vector by the angle θ in the clockwise direction; see Figure 8.1. Multiplying the canonical basis vectors of $\mathbb{R}^2$ by the diagonal matrix Σ, we have

$$\Sigma e_1 = \sigma_1 e_1 = \frac{6}{5} e_1, \quad \Sigma e_2 = \sigma_2 e_2 = \frac{1}{2} e_2;$$

that is, the canonical basis vectors are dilated by factors equal to the corresponding singular values. Finally, since multiplication by the matrix U rotates a vector by the angle φ in the counterclockwise direction,

$$\mathsf{U}(\sigma_1 e_1) = \sigma_1 u_1, \quad \mathsf{U}(\sigma_2 e_2) = \sigma_2 u_2.$$

The SVD therefore allows us to follow the action of the matrix A on a vector x step by step, as visualized in Figure 8.1.

Multiplication by an orthogonal matrix does not change either the 2-norm of a vector or the induced 2-norm of a matrix, implying that

$$\begin{aligned} \|\mathsf{A}\|_2 &= \max_{\|x\|_2=1} \|\mathsf{A}x\|_2 = \max_{\|\mathsf{V}^\mathsf{T} x\|_2=1} \|\mathsf{U}\Sigma\mathsf{V}^\mathsf{T} x\|_2 \\ &= \max_{\|x'\|_2=1} \|\mathsf{U}(\Sigma x')\|_2 = \max_{\|x'\|_2=1} \|\Sigma x'\|_2 \\ &= \|\Sigma\|_2 = \sigma_1, \end{aligned}$$

because the maximum stretching of a vector by the diagonal matrix Σ equals σ_1. Of course this is perfectly in line with the construction of the matrix Σ in the proof of Theorem 8.2.

The following theorem will play a central role in assessing the solvability of linear systems of equations.

Theorem 8.3. *The columns of an $m \times n$ matrix A are linearly independent if and only if $m \geq n$ and $\sigma_n > 0$.*

Proof. We begin by observing that $m \geq n$ is a necessary condition for linear independence: Since the columns of A are vectors in $\mathbb{R}^m$, if $n > m$, their number exceeds the dimension of the vector space $\mathbb{R}^m$; hence they cannot be linearly independent.

Let $n \leq m$. We prove first that $\sigma_n > 0$ is a necessary condition for linear independence: If $\sigma_n = 0$, then

$$\Sigma e_n = \begin{bmatrix} 0 \\ \vdots \\ \sigma_n \\ \vdots \\ 0 \end{bmatrix} = 0,$$

and, further,

$$\mathsf{A}\mathsf{V}e_n = \mathsf{U}\Sigma\mathsf{V}^\mathsf{T}\mathsf{V}e_n = \mathsf{U}\Sigma e_n = \mathsf{U}0 = 0. \tag{8.3}$$

Since $\mathsf{V}e_n \neq 0$, there is a linear combination of the columns of A with not all zero coefficients that gives the zero vector; thus the columns of A are not linearly independent.

To show that the condition is sufficient, assume that $\sigma_n > 0$. Then all singular values of A are positive, and the columns of Σ are linearly independent, because they are positive multiples of the first n canonical basis vectors of $\mathbb{R}^m$. If the columns of A are not linearly independent, there is a vector $z \in \mathbb{R}^n$, $z \neq 0$, such that $\mathsf{A}z = 0$. Writing A in terms of its SVD, this is equivalent to saying that there is vector $z \neq 0$ such that $\mathsf{U}\Sigma\mathsf{V}^\mathsf{T}z = 0$; hence $\mathsf{U}\Sigma z' = 0$ for some nonzero $z' = \mathsf{V}^\mathsf{T}z \in \mathbb{R}^n$. On the other hand, since U is invertible, this implies that $\Sigma z' = 0$, contradicting the linear independence of the columns of Σ. Therefore we conclude that the columns of A must be linearly independent. This completes the proof. $\square$

The SVD yields a simple formula for the inverse of an invertible square matrix.

Theorem 8.4. *An $m \times m$ matrix A is invertible if and only if its smallest singular value is positive. Moreover, if A is invertible and $\mathsf{A} = \mathsf{U}\Sigma\mathsf{V}^\mathsf{T}$ is its SVD, then*

$$\mathsf{A}^{-1} = \mathsf{V}\Sigma^{-1}\mathsf{U}^\mathsf{T},$$

where

$$\Sigma^{-1} = \begin{bmatrix} \sigma_1^{-1} & & \\ & \ddots & \\ & & \sigma_n^{-1} \end{bmatrix}.$$

Proof. If the $m \times m$ matrix A is invertible, its columns must be linearly independent, and therefore, from the previous theorem, the diagonal matrix Σ in its SVD has all positive diagonal entries. The inverse of the diagonal matrix Σ is a diagonal matrix with diagonal entries the reciprocals of the diagonal entries of Σ; hence U, Σ, and V^T are all invertible, and

$$\left(\mathsf{U}\Sigma\mathsf{V}^\mathsf{T}\right)^{-1} = \mathsf{V}\Sigma^{-1}\mathsf{U}^\mathsf{T} = \mathsf{A}^{-1}.$$

Vice versa, if the smallest, and hence all, singular values of A are positive, the matrix Σ is invertible, and $\mathsf{V}\Sigma^{-1}\mathsf{U}^\mathsf{T} = \mathsf{A}^{-1}$. □

8.3 ▪ The layered approach to the SVD, and lean SVD

The SVD not only is of great theoretical interest but also plays a key role when it is not feasible to keep all the entries of a matrix, and it becomes necessary to find an approximation that retains the most relevant traits of the matrix. Let r be the index of the smallest nonzero singular value of the $m \times n$ matrix A,

$$\sigma_1 \geq \sigma_2 \geq \cdots \geq \sigma_r > \sigma_{r+1} = \cdots = \sigma_{\min\{m,n\}} = 0.$$

Writing the SVD of the matrix A in terms of its left and right singular vectors,

$$\begin{aligned}
\mathsf{A} &= \begin{bmatrix} \vdots & & \vdots \\ u_1 & \cdots & u_m \\ \vdots & & \vdots \end{bmatrix} \begin{bmatrix} \sigma_1 & & & \\ & \ddots & & \\ & & \sigma_r & \\ & & & 0 \end{bmatrix} \begin{bmatrix} \cdots & v_1^\mathsf{T} & \cdots \\ & \vdots & \\ \cdots & v_n^\mathsf{T} & \cdots \end{bmatrix} \\
&= \begin{bmatrix} \vdots & & \vdots & & \vdots \\ \sigma_1 u_1 & \cdots & \sigma_r u_r & \cdots & 0 \\ \vdots & & \vdots & & \vdots \end{bmatrix} \begin{bmatrix} \cdots & v_1^\mathsf{T} & \cdots \\ & \vdots & \\ \cdots & v_n^\mathsf{T} & \cdots \end{bmatrix} \\
&= \sigma_1 u_1 v_1^\mathsf{T} + \sigma_2 u_2 v_2^\mathsf{T} + \cdots + \sigma_r u_r v_r^\mathsf{T},
\end{aligned}$$

yields an expression of the matrix A as a linear combination of r matrices of the form $u_j v_j^\mathsf{T}$,

$$\mathsf{A} = \sum_{j=1}^{r} \sigma_j u_j v_j^\mathsf{T}, \tag{8.4}$$

with coefficients the nonzero singular values. The triplets (u_j, v_j, σ_j), $1 \leq j \leq r$, form the *singular system* of the matrix, and they are sufficient to completely reconstruct the matrix.

Since the singular values are arranged in decreasing order, the matrices in the sum (8.4) are ranked in decreasing importance, with the components with more weight coming first, and the least important ones last. Therefore, if we approximate the matrix A by taking the first few terms of this sum, or, equivalently, by writing the sum of the matrices obtained from the first few singular triplets of the form (u_j, v_j, σ_j), we can hope that the most salient features of the matrix will be captured by the triplets corresponding to the largest singular values. In many applications, this process, referred to as a *low rank approximation* of the matrix A, is used to compress the information contained in the matrix. This reduction process is particularly useful when the dimensions of the original matrix are very large, or when the data represented in matrix form are noisy. If the sum extends over all nonzero singular values, it faithfully reproduces the matrix A.

Expressing a matrix as a linear combination of the matrices $u_j v_j^\mathsf{T}$ determined by its singular vectors shows that in order to recover the matrix A it is not necessary to know the singular vectors for indices larger than r. In matrix notation, the sum (8.4) can be written as

$$\begin{aligned}
\mathsf{A} &= \begin{bmatrix} \vdots & & \vdots \\ u_1 & \cdots & u_r \\ \vdots & & \vdots \end{bmatrix} \begin{bmatrix} \sigma_1 & & \\ & \ddots & \\ & & \sigma_r \end{bmatrix} \begin{bmatrix} \cdots & v_1^{\mathsf{T}} & \cdots \\ & \vdots & \\ \cdots & v_r^{\mathsf{T}} & \cdots \end{bmatrix} \\
&= \mathsf{U}_r \Sigma_r \mathsf{V}_r,
\end{aligned}$$

where

$$\mathsf{U}_r \in \mathbb{R}^{m \times r}, \quad \Sigma_r \in \mathbb{R}^{r \times r}, \quad \mathsf{V}_r \in \mathbb{R}^{n \times r}.$$

This factorization of the matrix is known as the economy-size, *lean, or thin SVD*. Although the lean SVD is sufficient to completely describe the action of the matrix, the additional information contained in the full SVD is very useful for analyzing the mapping properties of the matrix, as will be shown in the next chapter.

Problems

1. Using the SVD of the $m \times n$ matrix A, show that, if $n \leq m$, the matrix $\mathsf{A}^\mathsf{T}\mathsf{A}$ is invertible if and only if the columns of A are linearly independent.

2. Using the SVD of the $m \times n$ matrix A, show that, if $n \geq m$, the matrix $\mathsf{A}\mathsf{A}^\mathsf{T}$ is invertible if and only if the rows of A are linearly independent.

3. If $\mathsf{A} \in \mathbb{R}^{m \times n}$ with $m > n$ and all singular values are positive, is $\mathsf{A}^\mathsf{T}\mathsf{A}$ invertible? How about $\mathsf{A}\mathsf{A}^\mathsf{T}$? Use the SVD of A to justify your answers.

4. Assume that a 6×8 matrix A has the SVD

$$\mathsf{A} = \mathsf{U}\Sigma\mathsf{V}^\mathsf{T}$$

with singular values $21, 11, 6, 5, 3, 2$.

   (a) Show that the range of A is a subspace, determine its dimension, and find an orthonormal basis in terms of the SVD of A.

   (b) Show that the null space of A is a subspace, determine its dimension, and find an orthonormal basis in terms of the SVD of A.

   (c) Is the vector $z = 3\,v_6 - 11\,v_8$ in the null space of A? Here v_j denotes the jth column of the matrix V. Justify your answer.

5. Assume that a 10×10 matrix A has the SVD

$$\mathsf{A} = \mathsf{U}\Sigma\mathsf{V}^\mathsf{T}$$

with singular values $101, 61, 33, 22, 17, 11, 6, 3, 0, 0$.

   (a) What is the dimension of the range of A a subspace of? What is its dimension? Find an orthonormal basis in terms of the SVD of A.

   (b) What is the dimension of the null space of A? Find an orthonormal basis in terms of the SVD of A.

   (c) Is the matrix A invertible? If so, find its inverse in terms of its SVD .

Chapter 9

The Four Fundamental Subspaces

The singular value decomposition (SVD) reveals a lot of information about the properties of a matrix. In addition, as will be discussed in detail below, it is a very powerful tool to analyze systems of linear equations that can be expressed in the matrix-vector form as

$$\mathsf{A}x = b.$$

A full analysis of the solvability of a linear system needs to address both the existence and the uniqueness of the solution in terms of the coefficient matrix A and the right-hand side vector b. The interrelation between these two concepts is best understood in terms of four fundamental subspaces associated with the matrix A.

9.1 ▪ Subspace decompositions

We start with some results concerning the decomposition of vector spaces as sums of their subspaces. For the sake of simplicity, we restrict our discussion to the vector space $\mathbb{R}^n$. Analogous results hold for $\mathbb{C}^n$.

Definition 9.1. *Two subspaces $V, W \subset \mathbb{R}^n$ are* orthogonal *if*

$$v^{\mathsf{T}} w = 0 \textit{ for all } v \in V,\ w \in W.$$

To denote the orthogonality of the two subspaces we use the notation

$$V \perp W.$$

Because of the spanning property of basis vectors, to check if two subspaces are orthogonal it suffices to verify if every element of a basis of one is orthogonal to every element of a basis of the other.

Definition 9.2. *Let $V, W \subset \mathbb{R}^n$ be two subspaces.*

1. *If every $x \in \mathbb{R}^n$ can be decomposed as*

$$x = v + w, \quad v \in V, \quad w \in W, \tag{9.1}$$

then $\mathbb{R}^n$ is the sum of V and W, and we write

$$\mathbb{R}^n = V + W.$$

2. *If every* $x \in \mathbb{R}^n$ *admits a unique decomposition* (9.1), *then* $\mathbb{R}^n$ *is a* direct sum *of* V *and* W, *and we write*
$$\mathbb{R}^n = V \oplus W,$$

3. *If* $\mathbb{R}^n = V \oplus W$ *and, in addition, the subspaces* V *and* W *are mutually orthogonal, then* W *is the* orthocomplement *of* V, *and, vice versa,* V *is the orthocomplement of* W. *We denote orthocomplementarity by*
$$W = V^{\perp}, \quad V = W^{\perp}.$$

The uniqueness in part 2 means that if

$$x = v + w = v' + w'$$

for some vectors $v, v' \in V$ and $w, w' \in W$, then necessarily $v = v'$ and $w = w'$.

Example 11: Consider the subspaces of $\mathbb{R}^2$,

$$V = \operatorname{span}\{e_1, e_1 + e_2\}, \quad W = \operatorname{span}\{e_2\}.$$

Because every $x \in \mathbb{R}^2$ can be written as

$$x = \underbrace{x_1 e_1}_{=v} + \underbrace{x_2 e_2}_{=w} = v + w,$$

with $v \in V$, $w \in W$, we conclude that $\mathbb{R}^2 = V + W$. However, $\mathbb{R}^2$ is not the direct sum of V and W because the decomposition is not unique. In fact, we can write

$$x = \underbrace{x_1(e_1 + e_2)}_{=\tilde{v}} + \underbrace{(x_2 - x_1)e_2}_{=\tilde{w}} = \tilde{v} + \tilde{w},$$

which need not coincide with the previous decomposition.

Example 12: Consider the subspaces of $\mathbb{R}^3$,

$$V = \operatorname{span}\{e_1, e_2\}, \quad W = \operatorname{span}\{e_2 + e_3\}.$$

Every vector $x \in \mathbb{R}^3$ can be expressed as

$$x = x_1 e_1 + x_2 e_2 + x_3 e_3 = \underbrace{x_1 e_1 + (x_2 - x_3)e_2}_{=v} + \underbrace{x_3(e_2 + e_3)}_{=w} = v + w,$$

where $v \in V$, $w \in W$, so $\mathbb{R}^3 = V + W$. Further, it can be verified that the sum is direct. Let

$$v_1 = e_1, \quad v_2 = e_2, \quad w_1 = e_2 + e_3.$$

If a vector x has two representations,

$$x = v + w = v' + w', \quad v, v' \in V, \quad w, w' \in W,$$

we may write

$$x = \underbrace{\alpha_1 v_1 + \alpha_2 v_2}_{=v} + \underbrace{\alpha_3 w_1}_{=w} = \underbrace{\alpha_1' v_1 + \alpha_2' v_2}_{=v'} + \underbrace{\alpha_3' w_1}_{=w'},$$

implying that

$$(\alpha_1 - \alpha_1')v_1 + (\alpha_2 - \alpha_2')v_2 + (\alpha_3 - \alpha_3')w_1 = 0.$$

It follows from the linear independence of the vectors v_1, v_2, w_1 that

$$\alpha_j = \alpha_j', \quad 1 \le j \le 3,$$

proving the uniqueness of the representation. However, the two subspaces are not orthogonal, since

$$w_1^\mathsf{T} v_2 = 1 \neq 0.$$

Theorem 9.3. *Let V and W be two subspaces of $\mathbb{R}^n$ such that $\mathbb{R}^n = V + W$. Then*

$$\mathbb{R}^n = V \oplus W$$

if and only if

$$V \cap W = \{0\}.$$

In particular, if $\mathbb{R}^n = V + W$ and V and W are mutually orthogonal, then

$$\mathbb{R}^n = V \oplus W.$$

Proof. Assume that $V \cap W = \{0\}$. If $x \in \mathbb{R}^n$ has two representations,

$$x = v + w = v' + w',$$

then

$$v - v' = w - w',$$

implying that $v - v'$ and $w - w'$ are in the intersection of the two subspaces. Since $V \cap W = \{0\}$, then $v = v'$ and $w = w'$.

Conversely, if there is a vector $x \neq 0$ such that $x \in V \cap W$, then x has two different representations,

$$x = \underbrace{x}_{=v} + \underbrace{0}_{=w} = \underbrace{0}_{=v} + \underbrace{x}_{=w};$$

thus the sum is not direct.

Finally, we observe that the intersection of orthogonal subspaces can contain only the null vector. □

9.2 ▪ Four fundamental subspaces of a matrix

For the sake of definiteness, in the rest of this chapter we will restrict our discussion to matrices with real entries.

Definition 9.4. *Every matrix $\mathsf{A} \in \mathbb{R}^{m\times n}$ defines the following four fundamental subspaces:*

1. *Null space of* A*:*

$$\mathcal{N}(\mathsf{A}) = \{x \in \mathbb{R}^n \mid \mathsf{A}x = 0\} \subset \mathbb{R}^n.$$

2. *Range of* A*:*

$$\mathcal{R}(\mathsf{A}) = \{y \in \mathbb{R}^m \mid y = \mathsf{A}x \textit{ for some } x \in \mathbb{R}^n\} \subset \mathbb{R}^m.$$

3. *Null space of* A^T*:*

$$\mathcal{N}(\mathsf{A}^\mathsf{T}) = \{y \in \mathbb{R}^m \mid \mathsf{A}^\mathsf{T} y = 0\} \subset \mathbb{R}^m.$$

4. *Range of* A^T*:*

$$\mathcal{R}(\mathsf{A}^\mathsf{T}) = \{x \in \mathbb{R}^n \mid x = \mathsf{A}^\mathsf{T} y \textit{ for some } y \in \mathbb{R}^m\} \subset \mathbb{R}^n.$$

We have already shown that if

$$\mathsf{A} = \begin{bmatrix} \vdots & & \vdots \\ a_1 & \cdots & a_n \\ \vdots & & \vdots \end{bmatrix}, \quad a_j \in \mathbb{R}^m, \tag{9.2}$$

then

$$\mathcal{R}(\mathsf{A}) = \operatorname{span}\{a_1, \ldots, a_n\}.$$

This explains why the range of A is also called the *column space of* A.

Likewise, if we think of the matrix A in a row-wise fashion,

$$\mathsf{A} = \begin{bmatrix} \cdots & \alpha_1^\mathsf{T} & \cdots \\ & \vdots & \\ \cdots & \alpha_m^\mathsf{T} & \cdots \end{bmatrix}, \quad \alpha_j \in \mathbb{R}^n, \tag{9.3}$$

then

$$\mathcal{R}(\mathsf{A}^\mathsf{T}) = \operatorname{span}\{\alpha_1, \ldots, \alpha_m\},$$

which explains why the range of A^T is referred to as the *row space of* A.

Definition 9.5. *The* rank *of a matrix* A *is the dimension of its range* $\mathcal{R}(\mathsf{A})$*. Equivalently, the rank is the maximum number of linearly independent columns of* A*.*

The observation that

$$y^\mathsf{T}\mathsf{A} = 0 \text{ if and only if } \mathsf{A}^\mathsf{T} y = 0$$

justifies the name *left null space of* A for the subspace $\mathcal{N}(\mathsf{A}^\mathsf{T})$.

The following theorem establishes the pairwise orthogonality of the fundamental subspaces.

Theorem 9.6. *The null space of* A *is orthogonal to the row space of* A *with respect to the canonical inner product in* $\mathbb{R}^n$*, that is,*

$$\mathcal{N}(\mathsf{A}) \perp \mathcal{R}(\mathsf{A}^\mathsf{T}).$$

Likewise, the left null space of A *is orthogonal to the column space of* A*,*

$$\mathcal{N}(\mathsf{A}^\mathsf{T}) \perp \mathcal{R}(\mathsf{A}).$$

Proof. It follows from the definition of null space and the row-wise interpretation of matrix-vector product that every vector $x \in \mathcal{N}(\mathsf{A})$ is orthogonal to each row of the matrix A, or in terms of (9.3),

$$\alpha_j^\mathsf{T} x = 0, \quad 1 \le j \le m.$$

Since every vector $y \in \mathcal{R}(\mathsf{A}^\mathsf{T})$ can be written as a linear combination of the rows of A, it is orthogonal to every vector in $\mathcal{N}(\mathsf{A})$.

The same argument can be applied to prove the orthogonality of the left null space and the column space of the matrix A. $\square$

Another way to prove Theorem 9.6 is to observe that for $x \in \mathcal{N}(\mathsf{A})$ and $y \in \mathbb{R}^n$,

$$0 = (\mathsf{A}x)^\mathsf{T} y = x^\mathsf{T}(\mathsf{A}^\mathsf{T} y),$$

from which it follows that every vector x in the null space of A is orthogonal to every vector $\mathsf{A}^\mathsf{T} y \in \mathcal{R}(\mathsf{A}^\mathsf{T})$. A similar argument proves the second claim.

The next theorem relates the dimensions of the four fundamental subspaces to the number of nonzero singular values of the matrix. In the proof of the theorem, the columns u_j and v_j of the orthogonal matrices U, V in the SVD of A play a very important role.

Theorem 9.7. *Let the $m \times n$ matrix A have r positive singular values,*

$$\sigma_1 \geq \cdots \geq \sigma_r > \sigma_{r+1} = \cdots = \sigma_{\min(m,n)} = 0.$$

Then the following hold:

1. *The range of A has dimension r, and*

$$\mathcal{R}(\mathsf{A}) = \operatorname{span}\{u_1, \dots, u_r\}.$$

2. *The left null space of A has dimension $m - r$, and*

$$\mathcal{N}(\mathsf{A}^\mathsf{T}) = \operatorname{span}\{u_{r+1}, \dots, u_m\}.$$

3. *The row space of A has dimension r, and*

$$\mathcal{R}(\mathsf{A}^\mathsf{T}) = \operatorname{span}\{v_1, \dots, v_r\}.$$

4. *The null space of A has dimension $n - r$, and*

$$\mathcal{N}(\mathsf{A}) = \operatorname{span}\{v_{r+1}, \dots, v_n\}.$$

In particular, the rank of A, which by definition is maximum number of independent columns of the matrix, satisfies

$$\operatorname{rank}(\mathsf{A}) = r = \textit{number of positive singular values}.$$

Proof.

1. Every vector y in the range of A can be expressed in the form $y = \mathsf{A}x$ for some $x \in \mathbb{R}^n$. Substituting into A its layered SVD, $\mathsf{A} = \sum_{j=1}^r \sigma_j u_j v_j^\mathsf{T}$, we get

$$y = \sum_{j=1}^r \sigma_j u_j v_j^\mathsf{T} y = \sum_{j=1}^r \left(\sigma_j v_j^\mathsf{T} x\right) u_j;$$

thus

$$y \in \operatorname{span}\{u_1, \dots, u_r\}.$$

Conversely, every vector u_j, with $1 \leq j \leq r$, is in the range of $\mathsf{A} = \sum_{j=1}^r \sigma_j u_j v_j^\mathsf{T}$ since, by the mutual orthogonality of the vectors v_j,

$$u_j = \mathsf{A}\left(\frac{1}{\sigma_j} v_j\right).$$

2. It is straightforward to verify that the layered SVD of A^T is

$$\mathsf{A}^\mathsf{T} = \sum_{j=1}^{r} \sigma_j v_j u_j^\mathsf{T};$$

therefore

$$\mathsf{A}^\mathsf{T} y = \sum_{j=1}^{r} \left(\sigma_j u_j^\mathsf{T} y\right) v_j. \tag{9.4}$$

Using this formula, we conclude that a vector $y \in \mathbb{R}^m$ is in the left null space of A if and only if

$$\sum_{j=1}^{r} \left(\sigma_j u_j^\mathsf{T} y\right) v_j = 0.$$

The linear independence of the vectors v_j implies that

$$\sigma_1 u_1^\mathsf{T} y = \cdots = \sigma_r u_r^\mathsf{T} y = 0,$$

showing that y is orthogonal to $u_1, \ldots, u_r$; hence

$$y \in \operatorname{span}\{u_{r+1}, \ldots, u_m\}.$$

3. It follows from (9.4) and an argument analogous to that used to prove part 1 that the row space consists of linear combinations of vectors v_j.

4. A vector $x \in \mathbb{R}^n$ is in the null space of A if and only if

$$\sum_{j=1}^{r} \left(\sigma_j v_j^\mathsf{T} x\right) u_j = 0.$$

Following reasoning analogous to that used to prove part 2, we have

$$x \in \operatorname{span}\{v_{r+1}, \ldots, v_n\}. \qquad \square$$

Observe that the above theorem implies that the rank of a matrix can be defined equivalently as the dimension of its row space, or the maximum number of its linearly independent rows.

All matrices of the form

$$\mathsf{A} = v w^\mathsf{T}, \quad 0 \neq v \in \mathbb{R}^m, \quad 0 \neq w \in \mathbb{R}^n,$$

have rank 1 and are thus called *rank-*1 *matrices*. Moreover, one can check that the only nonzero singular value of such a matrix is $\sigma_1 = \|v\|\|w\|$.

We are now ready to formulate the results of Theorem 9.7 in terms of subspace decompositions.

Theorem 9.8. *The four fundamental subspaces associated with the* $m \times n$ *matrix* A *induce the subspace decompositions*

$$\mathbb{R}^n = \mathcal{N}(\mathsf{A}) \oplus \mathcal{R}(\mathsf{A}^\mathsf{T}),$$
$$\mathbb{R}^m = \mathcal{N}(\mathsf{A}^\mathsf{T}) \oplus \mathcal{R}(\mathsf{A}).$$

Moreover, the terms of the direct sums are pairwise mutually orthogonal; that is,

$$\mathcal{N}(\mathsf{A})^\perp = \mathcal{R}(\mathsf{A}^\mathsf{T}),$$
$$\mathcal{N}(\mathsf{A}^\mathsf{T})^\perp = \mathcal{R}(\mathsf{A}).$$

The SVD automatically provides orthonormal bases for the four fundamental subspaces. Symbolically, we partition the columns of the matrices U and V into blocks that are the orthonormal bases of the four fundamental subspaces as follows:

$$\mathsf{U} = \left[\begin{array}{c|c} \mathcal{R}(\mathsf{A}) & \mathcal{N}(\mathsf{A}^\mathsf{T}) \\ \leftarrow \quad r \quad \rightarrow & \leftarrow \quad m-r \quad \rightarrow \end{array}\right],$$

$$\mathsf{V} = \left[\begin{array}{c|c} \mathcal{R}(\mathsf{A}^\mathsf{T}) & \mathcal{N}(\mathsf{A}) \\ \leftarrow \quad r \quad \rightarrow & \leftarrow \quad n-r \quad \rightarrow \end{array}\right].$$

The results of this section are helpful for analyzing the properties of linear mappings defined in terms of matrix-vector products. We summarize some related results in the following theorems.

Theorem 9.9. *Consider the linear function*

$$f_\mathsf{A} : \mathbb{R}^n \to \mathbb{R}^m, \qquad x \mapsto \mathsf{A}x$$

associated with the matrix $\mathsf{A} \in \mathbb{R}^{m\times n}$. *Then the following hold:*

1. *If* $n > m$, *the function* f_A *is not injective.*
2. *If* $m > n$, *the function* f_A *is not surjective.*

Proof.

1. If $n > m$, the matrix A has at most $m = \min\{m, n\}$ nonzero singular values; hence, by Theorem 9.7, the dimension of its null space is at least $n - r \geq n - m > 0$. This implies that there is at least one vector $z \in \mathbb{R}^n$, $z \neq 0$, such that $\mathsf{A}z = 0$; hence the function defined by the matrix A is not injective since
$$f_\mathsf{A}(0) = 0 = f_\mathsf{A}(z).$$
2. If $m > n$, the number of nonzero singular values of A is at most $n = \min\{m, n\}$; hence the range of A is a subspace of $\mathbb{R}^m$ of dimension at most $n < m$. This implies that there is at least one vector in $\mathbb{R}^m$ that cannot be expressed as $\mathsf{A}x$ for some $x \in \mathbb{R}^n$. Therefore the function f_A is not surjective. □

In the finite dimensional case, the following result is a special instance of the *Fredholm alternative*.

Theorem 9.10 (Fredholm alternative). *Given a matrix* $\mathsf{A} \in \mathbb{R}^{m\times n}$ *and a vector* $b \in \mathbb{R}^m$, *then exactly one of the following statements must hold:*

1. *There exists a vector* $x \in \mathbb{R}^n$ *such that* $\mathsf{A}x = b$.
2. *There exists a vector* $y \in \mathcal{N}\left(\mathsf{A}^\mathsf{T}\right)$ *such that* $y^\mathsf{T} b \neq 0$.

Proof. The proof is based on the orthogonality of $\mathcal{R}(\mathsf{A})$ and $\mathcal{N}\left(\mathsf{A}^\mathsf{T}\right)$. Assuming that condition 1 holds, we show that condition 2 cannot hold. Indeed, if $b = \mathsf{A}x$ for some $x \in \mathbb{R}^n$, then $b \in \mathcal{R}(\mathsf{A})$, and therefore b is orthogonal to $\mathcal{N}\left(\mathsf{A}^\mathsf{T}\right)$. This implies that for all vectors $y \in \mathbb{R}^m$ such that $y^\mathsf{T}\mathsf{A} = 0$ (or $\mathsf{A}^\mathsf{T}y = 0$), $y^\mathsf{T}b = 0$; that is, condition 2 does not hold. This is equivalent to saying that b does not have any component along $\mathcal{N}\left(\mathsf{A}^\mathsf{T}\right)$.

Conversely, assume that condition 1 does not hold. We write b in terms of the basis $\{u_j\}$,

$$b = \sum_{j=1}^{r} \left(u_j^\mathsf{T} b\right) u_j + \sum_{k=r+1}^{m} \left(u_k^\mathsf{T} b\right) u_k,$$

where $r = \text{rank}(\mathsf{A})$. If b is not in $\mathcal{R}(\mathsf{A}) \neq \mathbb{R}^m$, then $r < m$ and, for some k, $r+1 \leq k \leq m$, $u_k^\mathsf{T} b \neq 0$. Since $u_k \neq 0$ is a basis vector of $\mathcal{N}\left(\mathsf{A}^\mathsf{T}\right)$, this means that condition 2 holds, completing the proof. □

The above theorem shows that in order to decide whether the equation $\mathsf{A}x = b$ has a solution, it is sufficient to check a finite number of conditions, namely whether or not b is orthogonal to the basis vectors of the subspace $\mathcal{N}(\mathsf{A}^\mathsf{T})$, which in some applications may be an easier task than showing the solvability otherwise.

Problems

1. Use the SVD and how it relates to the four fundamental subspaces to show that if a matrix A is invertible, its rows are linearly independent.

2. Use the four fundamental subspaces to show that if a matrix is non-square, then either its rows, or its columns, or both, are not linearly independent.

3. Show that if a matrix has linearly independent rows and linearly independent columns, then it is square and invertible.

4. Consider a matrix $\mathsf{A} = xx^\mathsf{T}$, where $x \in \mathbb{R}^n$, $x \neq 0$.

   (a) Show that $\text{rank}(\mathsf{A}) = 1$, and $\mathcal{R}(\mathsf{A}) = \text{span}\{x\}$.

   (b) Show that if $x \in \mathbb{R}^n$, $\|x\|_2 = 1$, every vector in the range of $\mathsf{B} = \mathsf{I}_n - \mathsf{A}$ is orthogonal to the vector x. Is the claim true if $\|x\|_2 \neq 1$?

   (c) Determine the rank of $\mathsf{B} = \mathsf{I}_n - \mathsf{A}$. Justify your answer in terms of the four fundamental subspaces of B.

5. Let V be a 5×3 matrix whose columns are orthonormal vectors $v_1,\ v_2,\ v_3$ in $\mathbb{R}^5$.

   (a) Show that the range of the matrix
   $$\mathsf{P}_V = \mathsf{V}\mathsf{V}^\mathsf{T}$$
   is the subspace V of $\mathbb{R}^5$ spanned by $v_1,\ v_2,\ v_3$.

   (b) Write an SVD for P_V that includes the vectors $v_1,\ v_2,\ v_3$. What are the singular values of P_V?

   (c) Show that
   $$\mathsf{P}_V\mathsf{P}_V = \mathsf{P}_V.$$

   (d) Show that the range of the matrix $\mathsf{I} - \mathsf{P}_V$ is the null space of P_V and, moreover,
   $$(\mathsf{I} - \mathsf{P}_V)\,(\mathsf{I} - \mathsf{P}_V) = \mathsf{I} - \mathsf{P}_V.$$

6. Consider a 4×2 matrix A and a 2×5 matrix B.

   (a) What are the possible dimensions of the null space of AB?

   (b) What are the possible dimensions of the range of AB?

   (c) Can the linear transformation defined by A be injective?

   (d) Can the linear transformation defined by B be surjective?

7. Is it possible to find a 3×3 matrix A such that the dimension of its range is equal to that of its null space? Explain.

8. Determine the dimensions of the four fundamental subspaces of the following three matrices:
$$\mathsf{A} = \begin{bmatrix} 1 & 1 & 1 \\ 1 & 1 & 1 \\ 1 & 1 & 1 \end{bmatrix}, \quad \mathsf{B} = \begin{bmatrix} 1 & 2 \\ 2 & 3 \\ 3 & 4 \end{bmatrix}, \quad \mathsf{C} = \begin{bmatrix} 5 & -4 & 3 \end{bmatrix}.$$

Chapter 10

Gaussian Elimination and Row Reduced Forms

One of the canonical uses of matrices is in the solution of linear systems of equations. In this section, we consider the classical method of solving linear systems by elimination of coefficients through linear combinations of individual equations, and we derive the corresponding LU matrix factorization.

Given a system of m equations in n unknowns of the form

$$\begin{aligned} a_{1,1}x_1 + a_{1,2}x_2 + \cdots + a_{1,n}x_n &= b_1, \\ a_{2,1}x_1 + a_{2,2}x_2 + \cdots + a_{2,n}x_n &= b_2, \\ &\cdots \\ a_{m,1}x_1 + a_{m,2}x_2 + \cdots + a_{m,n}x_n &= b_m, \end{aligned}$$

we collect the unknowns $x_1, \ldots, x_n$ into the vector $x \in \mathbb{R}^n$ and the known terms $b_1, \ldots, b_m$ into the vector $b \in \mathbb{R}^m$, and we define the *coefficient matrix*

$$\mathsf{A} = \begin{bmatrix} a_{1,1} & a_{1,2} & \cdots & a_{1,n-1} & a_{1,n} \\ a_{2,1} & a_{2,2} & \cdots & a_{2,n-1} & a_{2,n} \\ & & & & \\ a_{m,1} & a_{m,2} & \cdots & a_{m,n-1} & a_{m,n} \end{bmatrix}. \tag{10.1}$$

We can then write the linear system concisely as a matrix-vector equation of the form

$$\mathsf{A}x = b. \tag{10.2}$$

In this chapter, we separate the subscripts of the entries of the matrix A by a comma to emphasize that each row corresponds to an equation and each column to an unknown.

10.1 ▪ Square linear systems

Before discussing how to solve linear systems, we need to specify what we mean by solution.

Definition 10.1. *Any vector $x \in \mathbb{R}^n$ whose product with the matrix* A *is the vector b is a* solution *of the linear system* (10.2).

If the matrix A is square and invertible, then

$$\mathsf{A}^{-1}\mathsf{A}x = \mathsf{A}^{-1}b \;\Rightarrow\; x = \mathsf{A}^{-1}b;$$

that is, the solution exists, is unique, and is the product of the inverse of A and b.

In practice, even if a matrix is invertible, computing the inverse of A is not necessary for solving the associated linear system (10.2), and it is not advisable unless the structure of A makes its calculation very simple.

We begin by considering the cases where the solution of a square linear system can be computed easily. If A is an orthogonal matrix, its inverse is the transpose; therefore the linear system has a unique solution for any vector b. Moreover, the solution is the product of the transpose of the coefficient matrix and the right-hand side. Similarly, in the case where A is diagonal with all nonzero diagonal entries, the inverse exists and is the diagonal matrix whose diagonal entries are the reciprocals of the corresponding entries of the matrix A.

Next we consider square linear systems for which the solution, if it exists, can be computed in a straightforward way. If the matrix $\mathsf{A} = \mathsf{U}$ is square and upper triangular, that is, the entries of U below the diagonal vanish, the solution of the linear system (10.2), written componentwise as

$$\begin{aligned} u_{1,1}x_1 + u_{1,2}x_2 + \cdots + u_{1,n}x_n &= b_1, \\ u_{2,2}x_2 + \cdots + u_{2,n}x_n &= b_2, \\ \vdots \quad &\quad \vdots \\ u_{n,n}x_n &= b_n, \end{aligned}$$

if it exists, can be computed by a process known as *back substitution*: If $u_{n,n} \neq 0$, solving the last equation for x_n gives

$$x_n = \frac{b_n}{u_{n,n}}.$$

Substituting this expression for x_n in the next to last equation, and subtracting the x_n-dependent term from both sides, we have

$$u_{n-1,n-1}x_{n-1} = b_{n-1} - u_{n,n-1}x_n. \tag{10.3}$$

If $u_{n-1,n-1} \neq 0$, we can solve (10.3) for x_{n-1}:

$$x_{n-1} = \frac{b_{n-1} - u_{n,n-1}x_n}{u_{n-1,n-1}}.$$

We can continue this process recursively, as long as the diagonal entries of the matrix U are different from zero, until we arrive at the first equation, which we can solve for x_1. In this manner we determine all the components of the solution vector, from last to first. Therefore we conclude that the solution of a linear system whose coefficient matrix is square upper triangular with all nonzero diagonal entries exists and can be computed by the back substitution process.

If the matrix $\mathsf{A} = \mathsf{L}$ is square and lower triangular, that is, the entries above the diagonal vanish, the linear system is of the form

$$\begin{aligned} &\ell_{1,1}x_1 &&= b_1, \\ &\ell_{2,1}x_1 + \ell_{2,2}x_2 &&= b_2, \\ &\vdots &&\vdots \\ &\ell_{n,1}x_1 + \ell_{n,2}x_2 + \cdots + \ell_{n,n}x_n &&= b_n. \end{aligned}$$

As in the previous case, if $\ell_{1,1} \neq 0$, we can solve the first equation for x_1, substitute the value found in the second equation, and mimic the procedure for solving an upper triangular system, moving forward rather than backward. If the diagonal entries of L are all different from zero, this process, known as *forward substitution*, continues until we determine x_n.

We conclude that if the coefficient matrix of a linear system is either square upper or square lower triangular, and the solution exists, there is a straightforward method for computing it by forward or back substitution, respectively.

This observation can be used to design a strategy for computing the solution of a linear system with a generic coefficient matrix $\mathsf{A} \in \mathbb{R}^{n\times n}$, if it exists. We begin by observing that if the matrix A admits a factorization of the form

$$\mathsf{A} = \mathsf{LU}, \tag{10.4}$$

where L is lower triangular and U is upper triangular, substituting the right-hand side of (10.4) into (10.2) yields

$$\mathsf{LU}x = b. \tag{10.5}$$

If the diagonal entries of L and U are all different from zero, the solution of (10.5) exists and can be computed in two phases. Introducing $y = \mathsf{U}x$, we proceed as follows:

Phase 1: Solve

$$\mathsf{L}y = b$$

for y by forward substitution.

Phase 2: Solve

$$\mathsf{U}x = y$$

for x by back substitution.

Therefore, a linear system (10.2) with a solution can always be solved with forward and back substitution provided that we have have a factorization of the form (10.4) of its coefficient matrix. The identity (10.4) is called the *LU factorization*. Its properties and how to compute it are the topics of the next section.

10.2 ▪ Gaussian elimination and LU factorization

If a vector x solves the linear system (10.2), it also solves the linear system obtained by replacing one of the equations constituting the system with a linear combination of itself and the other equations in the system. This observation is the core of the process described below.

In our quest for an LU-factorization of A, we proceed by performing a sequence of *row operations* on A, replacing one row of the coefficient matrix at a time by a linear combination of the row itself and other rows to transform the matrix into upper triangular form. If the linear combination is chosen wisely, and certain conditions that will be specified later are satisfied, this process will introduce zeros below the main diagonal of A starting from the leftmost column and moving rightward.

When performing row operations to set to zero the entries below the main diagonal in the kth column, care must be taken not to overwrite previously introduced zeros with nonzero values.

In the following, we show step by step how the elimination of the nonzero entries is performed on an $m \times n$ matrix using appropriately constructed elimination matrices. When the procedure is applied to a nonsquare matrix, L is a square lower triangular matrix, while the matrix U, which has the same dimensions as the original coefficient matrix, will be as upper triangular as possible.

Before presenting the details of the LU factorization, we briefly summarize how multiplication by certain matrices is a way of performing elementary operations on the rows and columns of a given matrix. Let A be an $m \times n$ matrix.

1. Multiplying A from the left by an $m \times m$ diagonal matrix D_1 scales each row of A by the corresponding diagonal entry of D_1.

2. Multiplying A from the right by an $n \times n$ diagonal matrix D_2 scales each column of A by the corresponding diagonal entry of D_2.

3. Multiplying A from the left by an $m \times m$ identity with two swapped rows swaps the corresponding rows of A.

4. Multiplying A from the right by an $n \times n$ identity with two swapped columns swaps the corresponding columns of A.

10.2.1 ▪ Gaussian elimination with nonvanishing pivots

We are now ready to outline how the factorization of the $m \times n$ matrix A is carried out. In the first step of the LU factorization, consider a matrix of the form

$$\mathsf{E}_1 = \begin{bmatrix} 1 & & & \\ \ell_{2,1} & 1 & & \\ \vdots & & \ddots & \\ \ell_{m,1} & & & 1 \end{bmatrix} \in \mathbb{R}^{m \times m},$$

where all entries not on the diagonal or the first column vanish.

Writing the matrix A in terms of its rows,

$$\mathsf{A} = \begin{bmatrix} \cdots & \alpha_1^\mathsf{T} & \cdots \\ & \vdots & \\ \cdots & \alpha_m^\mathsf{T} & \cdots \end{bmatrix}, \qquad \alpha_j \in \mathbb{R}^n,$$

it is easy to see that multiplication from the left by the matrix E_1 yields

$$\mathsf{E}_1\mathsf{A} = \begin{bmatrix} \cdots & \alpha_1^\mathsf{T} & \cdots \\ \cdots & \alpha_2^\mathsf{T} + \ell_{2,1}\alpha_1^\mathsf{T} & \cdots \\ & \vdots & \\ \cdots & \alpha_m^\mathsf{T} + \ell_{m,1}\alpha_1^\mathsf{T} & \cdots \end{bmatrix}.$$

Our goal is to choose the coefficients $\ell_{j,1}$ so that the first component of each row vector below the first one vanishes, that is,

$$\mathsf{E}_1\mathsf{A} = \mathsf{A}_1 = \begin{bmatrix} a_{1,1} & a_{1,2} & \cdots & a_{1,n-1} & a_{1,n} \\ 0 & a_{2,2}^{(1)} & \cdots & a_{2,n-1}^{(1)} & a_{2,n}^{(1)} \\ \vdots & & & & \vdots \\ 0 & a_{m,2}^{(1)} & \cdots & a_{m,n-1}^{(1)} & a_{m,n}^{(1)} \end{bmatrix}.$$

The superscript in parentheses indicates that the entry has undergone one transformation due to the multiplication by E_1. To ensure that the entry in position $(j, 1)$ of the transformed matrix vanishes, $\ell_{j,1}$ must satisfy

$$\ell_{j,1}a_{1,1} + a_{j,1} = 0 \Rightarrow \ell_{j,1} = -\frac{a_{j,1}}{a_{1,1}}, \quad 2 \le j \le m,$$

provided that $a_{1,1} \neq 0$, thus providing a way to determine E_1 and, as it will turn out, the first column of L.

The effect of multiplying a vector by the matrix E_1 is to add to its jth entry $\ell_{j,1}$ times its first entry; thus we deduce that the inverse of the matrix E_1 must undo the action of E_1 by subtracting what was added. In other words, we conclude that

$$\mathsf{E}_1^{-1} = \begin{bmatrix} 1 & & & \\ -\ell_{2,1} & 1 & & \\ \vdots & & \ddots & \\ -\ell_{n,1} & & & 1 \end{bmatrix}.$$

It is straightforward to verify that $\mathsf{E}_1\mathsf{E}_1^{-1} = \mathsf{I}$. We will make use of the particular structure of the inverse later.

In the second step, we introduce zeros in the entries of the second column of A_1 below the main diagonal by multiplying A_1 from the left by the elementary matrix

$$\mathsf{E}_2 = \begin{bmatrix} 1 & & & & \\ & 1 & & & \\ & \ell_{3,2} & 1 & & \\ & \vdots & & \ddots & \\ & \ell_{m,2} & & & 1 \end{bmatrix} \in \mathbb{R}^{m \times m}.$$

We choose the coefficients in the second column of E_2 so that

$$\mathsf{E}_2\mathsf{E}_1\mathsf{A} = \mathsf{E}_2\mathsf{A}_1 = \begin{bmatrix} a_{1,1} & a_{1,2} & a_{1,3} & \cdots & a_{1,n-1} & a_{1,n} \\ 0 & a_{2,2}^{(1)} & a_{2,3}^{(1)} & \cdots & a_{2,n-1}^{(1)} & a_{2,n}^{(1)} \\ 0 & 0 & a_{3,3}^{(2)} & \cdots & a_{3,n-1}^{(2)} & a_{3,n}^{(2)} \\ \vdots & & & & & \vdots \\ 0 & 0 & a_{m,3}^{(2)} & \cdots & a_{m,n-1}^{(2)} & a_{m,n}^{(2)} \end{bmatrix},$$

which is tantamount to requiring that

$$\ell_{j,2}a_{2,2}^{(1)} + a_{j,2}^{(1)} = 0 \Rightarrow \ell_{j,2} = -\frac{a_{j,2}^{(1)}}{a_{2,2}^{(1)}}, \quad 3 \leq j \leq m,$$

thus providing a way to determine E_2 and the second column of L, provided that $a_{2,2}^{(1)} \neq 0$.

It can be verified that the matrix E_2 is invertible, and that its inverse is

$$\mathsf{E}_2^{-1} = \begin{bmatrix} 1 & & & & \\ & 1 & & & \\ & -\ell_{3,2} & 1 & & \\ & \vdots & & \ddots & \\ & -\ell_{m,2} & & & 1 \end{bmatrix}.$$

We continue the process recursively, constructing at the kth step the matrix $\mathsf{E}_k \in \mathbb{R}^{m\times m}$ such that, if $a_{k,k}^{(k-1)} \neq 0$, the entries under the kth diagonal entry of $\mathsf{E}_k\mathsf{A}_{k-1}$ are set to zero. If

$m \leq n$, after $m-1$ steps we obtain

$$\mathsf{E}_{m-1}\mathsf{E}_{m-2}\dots\mathsf{E}_2\mathsf{E}_1\mathsf{A} = \mathsf{U} = \begin{bmatrix} a_{1,1} & a_{1,2} & a_{1,3} & \cdots & a_{1,m} & \cdots & a_{1,n} \\ & a_{2,2}^{(1)} & a_{2,3}^{(1)} & \cdots & a_{2,m}^{(1)} & \cdots & a_{2,n}^{(1)} \\ & & a_{3,3}^{(2)} & \cdots & a_{3,m}^{(2)} & \cdots & a_{3,n}^{(2)} \\ & & & \ddots & \vdots & & \vdots \\ & & & & a_{m,m}^{(m-1)} & \cdots & a_{m,n}^{(m-1)} \end{bmatrix},$$

where the right-hand side is an upper triangular matrix with dimensions equal to those of A. Likewise, if $m > n$, after n steps we have an upper triangular matrix with zeros in the last $m-n$ rows,

$$\mathsf{E}_n\mathsf{E}_{n-1}\dots\mathsf{E}_2\mathsf{E}_1\mathsf{A} = \mathsf{U} = \left[\begin{array}{cccc} a_{1,1} & a_{1,2} & \cdots & a_{1,n} \\ & a_{2,2}^{(1)} & & a_{2,n}^{(1)} \\ & & \ddots & \vdots \\ & & & a_{n,n}^{(n)} \\ \hline & \mathsf{O}_{(m-n)\times n} & & \end{array}\right].$$

For the sake of definiteness, let us assume that $m \leq n$ and all matrices E_j are well defined. It follows from the invertibility of the matrices E_j that

$$\mathsf{A} = \mathsf{E}_1^{-1}\mathsf{E}_2^{-1}\dots\mathsf{E}_{m-2}^{-1}\mathsf{E}_{m-1}^{-1}\mathsf{U}.$$

Because the matrix E_j differs from the identity only in the entries in the jth column below the main diagonal, it can be verified directly that products of the form

$$\mathsf{E}_j^{-1}\mathsf{E}_{j+k}^{-1}$$

are equal to a unit matrix with the $(j+k)$th column replaced by the $(j+k)$th column of E_{j+k}^{-1}, and the jth column replaced by the jth column of E_j^{-1}. More generally, we have

$$\mathsf{E}_1^{-1}\mathsf{E}_2^{-1}\dots\mathsf{E}_{m-2}^{-1}\mathsf{E}_{m-1}^{-1} = \mathsf{L} = \begin{bmatrix} 1 & & & & \\ -\ell_{2,1} & 1 & & & \\ -\ell_{3,1} & -\ell_{3,2} & 1 & & \\ \vdots & & & \ddots & \\ -\ell_{m,1} & -\ell_{m,2} & & -\ell_{m,m-1} & 1 \end{bmatrix}; \tag{10.6}$$

hence

$$\mathsf{A} = \mathsf{LU}, \tag{10.7}$$

where U is an upper triangular and L is a lower triangular matrix.

The decomposition (10.7) is called the LU factorization of the matrix A.

Definition 10.2. *The nonvanishing diagonal elements*

$$a_{1,1},\ a_{2,2}^{(1)},\ \dots,\ a_{m-1,m-1}^{(m-2)}$$

are the pivots *of the matrix* A.

The procedure outlined above for determining the matrices L and U is called *Gaussian elimination*.

10.2.2 ▪ Partial pivoting

If, in the course of Gaussian elimination, the kth pivot vanishes, the process as outlined above falls apart, as the division by the vanishing pivot, required for computing the factors $\ell_{j,k}$, cannot be performed. Since the kth row of the matrix contains the coefficients of the kth linear equation, exchanging the order of the equations does not change whether or not there is a solution, or the possible solution itself. Therefore, if at least one of the last $m-k-1$ entries of the kth column, for example the one in the jth row, with $j > k$, is different from zero, we can swap the kth and jth rows and continue the elimination, taking care, at the end of the elimination, to reorder the entries of the right-hand vector b accordingly. To understand how row swapping alters the Gaussian elimination algorithm and the matrix factorization, we need to look at the details of the process.

In the case where $\mathsf{A} \in \mathbb{R}^{m\times n}$, $m \leq n$, and $\mathsf{P} \in \mathbb{R}^{m\times m}$ is the matrix obtained by swapping the ith and jth rows of an $m \times m$ identity matrix,

$$\mathsf{PA} = \text{matrix A with rows } i \text{ and } j \text{ swapped},$$

while if $\mathsf{P} \in \mathbb{R}^{n\times n}$ is a matrix obtained by swapping rows i and j of the identity matrix,

$$\mathsf{AP}^\mathsf{T} = \text{matrix A with columns } i \text{ and } j \text{ swapped}.$$

If A is a square matrix, the product PAP^T, referred to as the symmetric swap, is the matrix obtained by swapping the ith and jth rows and columns of A. Moreover,

$$\mathsf{PP}^\mathsf{T} = \mathsf{P}^\mathsf{T}\mathsf{P} = \mathsf{I}.$$

If $\mathsf{E}_1 \in \mathbb{R}^{m\times m}$ is a matrix of the form

$$\mathsf{E}_1 = \begin{bmatrix} 1 & & & \\ w_2 & 1 & & \\ \vdots & & \ddots & \\ w_m & & & 1 \end{bmatrix},$$

and P is the matrix swapping the rows i and j, where $i, j, > 1$, then

$$\mathsf{PE}_1\mathsf{P}^\mathsf{T} = \widetilde{\mathsf{E}}_1 = \begin{bmatrix} 1 & & & \\ \widetilde{w}_2 & 1 & & \\ \vdots & & \ddots & \\ \widetilde{w}_m & & & 1 \end{bmatrix},$$

where $\widetilde{w}_\ell = w_\ell$ for $\ell \neq i, j$, while $\widetilde{w}_i = w_j$ and $\widetilde{w}_j = w_i$. Hence,

$$\mathsf{PE}_1 = \widetilde{\mathsf{E}}_1\mathsf{P}; \tag{10.8}$$

that is, while the matrix-matrix product of P and E_1 does not commute, we may move the matrix P on the left of E_1 to the right side of the elementary matrix by replacing E_1 with $\widetilde{\mathsf{E}}_1$, an operation that preserves the pattern of nonzero entries.

More generally, if we have a matrix E of the form

$$\mathsf{E}_\ell = \begin{bmatrix} 1 & & & & \\ & \ddots & & & \\ & & 1 & & \\ & & w_{\ell+1} & \ddots & \\ & & \vdots & & \ddots \\ & & w_m & & & 1 \end{bmatrix} \in \mathbb{R}^{m\times m},$$

and $\mathsf{P} \in \mathbb{R}^{m\times m}$ is a matrix swapping the ith and jth rows, with $i,j > \ell$, then (10.8) holds with $\mathsf{E} = \mathsf{E}_\ell$, and $\widetilde{E}_\ell$ is otherwise identical to E_ℓ, with only the (i,ℓ)th and (j,ℓ)th entries in the nontrivial column swapped.

We are now ready to return to Gaussian elimination. Assume that $a_{1,1} = 0$, but $a_{j,1} \neq 0$ for some $j > 1$, or that $a_{1,1} \neq 0$, but $|a_{1,1}| \ll |a_{j,1}|$ for some j, causing potentially numerical problems when computing the ratio $a_{j,1}/a_{1,1}$ in finite precision arithmetic. In that case, instead of using $a_{1,1}$ as a pivot, we first modify the matrix A by swapping the first row with a row containing the entry of the first column of largest absolute value. Denote by $\mathsf{P}_1 \in \mathbb{R}^{m\times m}$ the corresponding swap matrix. Proceeding as before, we then apply the elimination process to the matrix $\mathsf{P}_1\mathsf{A}$, obtaining

$$\mathsf{E}_1\mathsf{P}_1\mathsf{A} = \mathsf{A}_1,$$

where $\mathsf{A}_1 \in \mathbb{R}^{m\times n}$ is a matrix with zero entries in the first column below the diagonal.

If the newly computed diagonal entry $a_{2,2}^{(1)} = 0$, and at least one entry $a_{j,2} \neq 0$ for $j \geq 2$, or, alternatively, $|a_{2,2}| \ll |a_{j,2}|$ for some $j > 2$, we can perform another row permutation to replace $a_{2,2}^{(1)}$, which is equivalent to multiplying A_1 by a permutation matrix P_2, and subsequently perform the elimination on the matrix $\mathsf{P}_2\mathsf{A}_1$.

This yields

$$\mathsf{E}_2\mathsf{P}_2\mathsf{E}_1\mathsf{P}_1\mathsf{A} = \mathsf{A}_2,$$

which, in view of (10.8), can be written as

$$\mathsf{E}_2\widetilde{\mathsf{E}}_1(\mathsf{P}_2\mathsf{P}_1)\mathsf{A} = \mathsf{A}_2.$$

The rest of the elimination process proceeds in the same manner. In general, at the ℓth step, do the following:

1. Permute the rows of the ℓth column of A_ℓ to select a pivot entry with large absolute value, if possible. This is equivalent to multiplying A_ℓ from the left by a row swapping matrix P_ℓ.

2. Eliminate the entries below the pivot, which is equivalent to a left multiplication by the elimination matrix matrix E_ℓ.

3. Commute the row swapping matrix P_ℓ and all the previous elimination matrices suitably modified.

At the end of the process we arrive at a factorization of the form

$$\left(\mathsf{E}_{m-1}\widetilde{\mathsf{E}}_{m-2}\cdots\widetilde{\mathsf{E}}_1\right)(\mathsf{P}_{m-1}\cdots\mathsf{P}_1)\,\mathsf{A} = \mathsf{U}. \tag{10.9}$$

Observing that the product of row-swapping matrices is a row permutation

$$\mathsf{P} = \mathsf{P}_{m-1}\cdots\mathsf{P}_1,$$

we see that (10.9) is equivalent to

$$\mathsf{PA} = \mathsf{LU},$$

where L is the lower triangular matrix $\widetilde{E}_1^{-1}\ldots\widetilde{E}_{m-1}^{-1}$, thus showing that the lower and upper triangular matrices determined in this manner constitute an LU factorization of a row-permuted version of A.

In the discussion above, we assume that the jth column of the matrix A_j contains a nonzero entry $a_{kj} \neq 0$ with $k \geq j$ that can be brought into pivot position through row swaps. If all the entries in or below the pivot position vanish, the elimination process moves to the next column. Notice that when a column is skipped, there is no need for row operations, so the product of the E matrices remains unaltered. In this case the final product of Gaussian elimination is a *step upper triangular matrix* of the form

$$
\mathsf{U} = \begin{bmatrix}
\mathbf{u_{1,1}} & u_{1,2} & u_{1,3} & u_{1,4} & \cdots & \cdots & u_{1,n-2} & u_{1,n-1} & u_{1,n} \\
0 & \mathbf{u_{2,2}} & u_{2,3} & u_{2,4} & \cdots & \cdots & u_{2,n-2} & u_{2,n-1} & u_{2,n} \\
0 & 0 & 0 & \mathbf{u_{3,4}} & \cdots & \cdots & u_{3,n-2} & u_{3,n-1} & u_{3,n} \\
 & & & & & & \vdots & \vdots & \vdots \\
 & & & & \cdots & \cdots & \mathbf{u_{k,n-2}} & u_{k,n-1} & u_{k,n} \\
 & & & & \cdots & \cdots & 0 & 0 & \mathbf{u_{r,n}} \\
 & & & & & & \vdots & \vdots & \vdots \\
 & & & & \cdots & \cdots & 0 & 0 & 0
\end{bmatrix}. \tag{10.10}
$$

In this case the pivots are the first nonzero entries of each row, indicated in boldface in the matrix above.

10.3 ▪ Row reduced echelon form

Gaussian elimination transforms the matrix A into upper triangular, or step upper triangular, form via a sequence of row operations. We can continue to perform row operations to further reduce the upper triangular matrix in the following way.

After identifying the last pivot, rescale the corresponding row by the reciprocal of the pivot value, and then add multiples of that row to the previous rows to introduce zeros above the pivot, which now is equal to 1. Since all the entries in the row to the left of the pivot are equal to zero, this sequence of operations preserves the upper triangular structure of the matrix. Once this elimination above the pivot has been completed, we move to the pivot in the row above and repeat the process until we reach the first pivot.

For example, the matrix U in (10.10) after the described elimination process assumes the form

$$
\mathsf{A}_{\mathrm{RREF}} = \begin{bmatrix}
\mathbf{1} & 0 & u'_{1,3} & 0 & \cdots & \cdots & 0 & u'_{1,n-1} & 0 \\
0 & \mathbf{1} & u'_{2,3} & 0 & \cdots & \cdots & 0 & u'_{2,n-1} & 0 \\
0 & 0 & 0 & \mathbf{1} & \cdots & \cdots & 0 & u'_{3,n-1} & 0 \\
 & & & & & & & & \\
 & & & & \cdots & \cdots & \mathbf{1} & u_{k,n-1} & 0 \\
 & & & & \cdots & \cdots & 0 & 0 & \mathbf{1} \\
 & & & & & & & & \\
 & & & & \cdots & \cdots & 0 & 0 & 0
\end{bmatrix}, \tag{10.11}
$$

where the 1s are now replacing the pivots.

A matrix that has undergone the two sweeps of Gaussian elimination is said to be in *row reduced echelon form* (RREF). If the entries of the right-hand side of the underlying linear system are subjected to the same operations applied to the coefficient matrix A, the linear system with coefficient matrix (10.13) and the transformed right-hand side has the same solution as the original one.

We can now prove the following important result concerning the rank of a matrix in RREF.

Theorem 10.3. *The dimension r of the range of* $\mathsf{A}_{\mathrm{RREF}}$ *is equal to the number of pivots, and the dimension of its null space is $n-r$. Moreover, the columns of the matrix* A *corresponding to the pivot columns of* $\mathsf{A}_{\mathrm{RREF}}$ *are a basis for the range of* A*. The nonpivot columns of the matrix* $\mathsf{A}_{\mathrm{RREF}}$ *can be used to find a basis for the null space of* A.

Proof. If the matrix $\mathsf{A}_{\mathrm{RREF}}$ has r pivots, the pivot columns are the standard basis vectors $e_1, \ldots, e_r$ of $\mathbb{R}^m$, and the last $m - r$ rows of $\mathsf{A}_{\mathrm{RREF}}$ are zero rows. Therefore, any vector of the form $y = \mathsf{A}_{\mathrm{RREF}} x \in \mathbb{R}^m$ has the last $m - r$ components equal to zero.

On the other hand, if we choose x such that it has nonzero components only in the position corresponding to pivot indices, then $y = \mathsf{A}x$ is a linear combination of the pivot columns. This reasoning shows that the range of $\mathsf{A}_{\mathrm{RREF}}$,

$$\mathcal{R}(\mathsf{A}_{\mathrm{RREF}}) = \left\{ y = \begin{bmatrix} z \\ 0 \end{bmatrix} \mid z \in \mathbb{R}^r \right\}, \tag{10.12}$$

is of dimension r.

To prove that the columns of A corresponding to the pivot columns of $\mathsf{A}_{\mathrm{RREF}}$ are a basis for the range of A, observe that since the row operations are reversible, we may write $\mathsf{FA} = \mathsf{A}_{\mathrm{RREF}}$, where F is an invertible matrix encoding the row operations. Therefore, if y is in the range of A,

$$y = \mathsf{A}x = \mathsf{F}^{-1}\big(\mathsf{A}_{\mathrm{RREF}} x\big).$$

If follows from (10.12) that

$$\mathsf{A}_{\mathrm{RREF}} x = \sum_{j=1}^{r} z_j e_j;$$

thus,

$$y = \mathsf{F}^{-1}\big(\sum_{j=1}^{r} z_j e_j\big) = \sum_{j=1}^{r} z_j F^{-1} e_j = \sum_{j=1}^{r} z_j a_{p_j},$$

where a_{p_j} is the column of A corresponding to the pivot column e_j of $\mathsf{A}_{\mathrm{RREF}}$. This proves that the columns of A corresponding to the pivot columns of $\mathsf{A}_{\mathrm{RREF}}$ span the range of A.

Their linear independence follows from the linear independence of $e_1, \ldots, e_r$. If

$$\sum_{j=1}^{r} z_j a_{p_j} = 0,$$

multiplying both sides by the matrix F shows that $z = 0$, implying linear independence of the columns of A. In particular, we conclude that

$$\dim\big(\mathcal{R}(\mathsf{A})\big) = r.$$

Since $\mathsf{A}_{\mathrm{RREF}} = \mathsf{FA}$ and F is invertible, the null spaces of A and $\mathsf{A}_{\mathrm{RREF}}$ coincide because

$$\mathsf{A}_{\mathrm{RREF}} x = \mathsf{FA}x = 0 \text{ if and only if } \mathsf{A}x = 0.$$

Thus a basis for the null space of $\mathsf{A}_{\mathrm{RREF}}$ is also a basis for the null space of A, and from the properties of the four fundamental subspaces it follows that the null spaces of $\mathsf{A}_{\mathrm{RREF}}$ and A both have dimension $n - r$.

To find a basis for the null space of $\mathsf{A}_{\mathrm{RREF}}$ we assume for simplicity that its first r columns are pivot columns. This can be achieved by multiplying $\mathsf{A}_{\mathrm{RREF}}$ by a column permutation matrix P^{T}, and noticing that this can be thought of as a change of variables,

$$\mathsf{A}_{\mathrm{RREF}} z = \big(\mathsf{A}_{\mathrm{RREF}} \mathsf{P}^{\mathsf{T}}\big)\big(\mathsf{P}z\big) = \widetilde{\mathsf{A}}_{\mathrm{RREF}} \widetilde{z}, \quad \widetilde{z} = \mathsf{P}z.$$

We partition the matrix $\mathsf{A}_{\mathrm{RREF}}$ as

$$\mathsf{A}_{\mathrm{RREF}} = \left[\begin{array}{c|ccc} & \vdots & & \vdots \\ \mathsf{I}_r & b_1 & \cdots & b_{n-r} \\ & \vdots & & \vdots \\ \hline \mathsf{O}_{(m-r)\times r} & & \mathsf{O}_{(m-r)\times(n-r)} & \end{array}\right] = \begin{bmatrix} \mathsf{I} & B \\ 0 & 0 \end{bmatrix}, \quad b_j \in \mathbb{R}^r, \tag{10.13}$$

and let $z_j \in \mathbb{R}^n$ be a vector of the form

$$z_j = \begin{bmatrix} \widehat{z}_j \\ e_j \end{bmatrix}, \quad \widehat{z}_j \in \mathbb{R}^r, \quad e_j \in \mathbb{R}^{n-r},$$

where e_j is the jth canonical coordinate vector in $\mathbb{R}^{n-r}$. The vector z_j is in the null space of $\mathsf{A}_{\mathrm{RREF}}$ if and only if

$$\mathsf{A}_{\mathrm{RREF}} z_j = \begin{bmatrix} \widehat{z}_j + b_j \\ 0 \end{bmatrix} = 0, \text{ or } z_j = \begin{bmatrix} -b_j \\ e_j \end{bmatrix}.$$

The vectors $z_1, \ldots, z_{n-r}$ determined in this manner are in the null space of $\mathsf{A}_{\mathrm{RREF}}$, and since they are linearly independent, they are a basis for the $(n-r)$-dimensional subspace $\mathcal{N}(\mathsf{A}_{\mathrm{RREF}}) = \mathcal{N}(\mathsf{A})$. We end the process by permuting the entries of the basis vectors by multiplying with the row permutation P. This completes the proof. □

The construction of the basis of the null space of A shows that every vector w in the null space can be written as

$$w = \begin{bmatrix} -\mathsf{B} \\ \mathsf{I}_{n-r} \end{bmatrix} v, \quad v \in \mathbb{R}^{n-r}; \tag{10.14}$$

thus the components of v are the coordinates of w in the basis $\{z_1, \ldots, z_{n-r}\}$.

We finish this section with the following observation.

Theorem 10.4. *If the $n \times n$ matrix A is invertible, then $\mathsf{A}_{\mathrm{RREF}}$ is the identity matrix.*

Proof. Because the range of an invertible $n \times n$ matrix has dimension n, every column is a pivot column; hence the row reduced form of the matrix is the identity. □

10.4 ▪ The general solution of a linear system

We are now ready to characterize all possible solutions of a linear system (10.2).

Theorem 10.5. *If a linear system* (10.2) *admits a solution $x_p \in \mathcal{R}(\mathsf{A}^\mathsf{T})$, then all vectors of the form*

$$x = x_p + w$$

with $w \in \mathcal{N}(\mathsf{A})$ are also solutions.

Proof. Assume that the matrix A has r pivots and that they appear in the first r columns of A. Denote the block of pivot columns by A_1, and partition the matrix A as

$$\mathsf{A} = \left[\, \mathsf{A}_1 \mid \mathsf{A}_2 \,\right], \quad \mathsf{A}_1 \in \mathbb{R}^{m\times r}, \quad \mathsf{A}_2 \in \mathbb{R}^{m\times(n-r)}.$$

Since by assumption the linear system admits a solution, the vector b is in the range of A; hence it can be expressed as a unique linear combination of the pivot columns, because they are a basis for $\mathcal{R}(\mathsf{A})$,

$$b = \mathsf{A}_1 z, \quad z \in \mathbb{R}^r.$$

If x^* is the vector

$$x^* = \begin{bmatrix} z \\ 0 \end{bmatrix} \in \mathbb{R}^n,$$

then

$$\mathsf{A}x^* = \mathsf{A}_1 z + \mathsf{A}_2 0 = b;$$

that is, x^* is a solution of the linear system. To find the particular solution x_p in the range of A^T, observe that if w is in the null space of A,

$$\mathsf{A}(x^* + w) = \mathsf{A}x^* + \mathsf{A}w = \mathsf{A}x^* = b;$$

therefore $x = x^* + w$ is also a solution.

To find the solution $x_p = x^* + w$ that is orthogonal to all basis vectors of the null space of A, and hence in the range of A^T, we write w using the representation (10.14) and require that

$$\begin{bmatrix} -\mathsf{B} \\ \mathsf{I}_{n-r} \end{bmatrix}^\mathsf{T} \left(\begin{bmatrix} z \\ 0 \end{bmatrix} + \begin{bmatrix} -\mathsf{B} \\ \mathsf{I}_{n-r} \end{bmatrix} v \right) = 0,$$

or, equivalently,

$$(\mathsf{I}_{n-r} + \mathsf{B}^\mathsf{T}\mathsf{B})v = \mathsf{B}^\mathsf{T} z. \tag{10.15}$$

Denoting by $\sigma_1 \geq \cdots \geq \sigma_{n-r} \geq 0$ the singular values of the matrix B, it follows that the singular values of the matrix $\mathsf{I}_{n-r} + \mathsf{B}^\mathsf{T}\mathsf{B}$ are $1 + \sigma_1^2 \geq \cdots \geq 1 + \sigma_{n-r}^2 \geq 1$. Therefore, the matrix is invertible, and the linear system (10.15) has a unique solution $v \in \mathbb{R}^{n-r}$.

The vector

$$x_p = \begin{bmatrix} z - \mathsf{B}v \\ -v \end{bmatrix}$$

is the *particular solution* in $\mathcal{R}(\mathsf{A}^\mathsf{T}) = \mathcal{N}(\mathsf{A})^\perp$. Moreover, every vector of the form

$$x = x_p + w \tag{10.16}$$

with $w \in \mathcal{N}(\mathsf{A})$ is also a solution. The expression (10.16) is referred to as the *general solution* of the linear system. If A is invertible, then $x = x_p$. □

Next we present an example where we illustrate how to use these concepts. We begin by remarking that it is not necessary to compute the RREF of the coefficient matrix to analyze the solvability of a linear system. Moreover, any particular solution can be used in the general solution—not necessarily the one that is orthogonal to the null space of the matrix.

Consider the matrix $\mathsf{A} \in \mathbb{R}^{6\times 7}$ and the vector $b \in \mathbb{R}^6$,

$$\mathsf{A} = \begin{bmatrix} 1 & 3 & 3 & 1 & 2 & 3 & 0 \\ 0 & 3 & 4 & 2 & 0 & 1 & 1 \\ 0 & 0 & 2 & -1 & -2 & 0 & 2 \\ 0 & 0 & 0 & 0 & -1 & 3 & 0 \\ 0 & 0 & 0 & 0 & 0 & 5 & -2 \\ 0 & 0 & 0 & 0 & 0 & 0 & 0 \end{bmatrix}, \quad b = \begin{bmatrix} 1 \\ 1 \\ 0 \\ 2 \\ 3 \\ 0 \end{bmatrix}.$$

The matrix A is already in step upper triangular form, and the linear system $\mathsf{A}x = b$ has a solution because the zero in the sixth entry of the vector b matches the row of zeros in A.

To compute a particular solution x^*, we need to find the vector whose nonzero entries are the coefficients of the linear combination of the pivot columns that gives the vector b. In this case we set to zero the fourth and seventh entries of x^* associated with nonpivot columns, and we solve the linear system

$$\begin{bmatrix} 1 & 3 & 3 & 2 & 0 \\ 0 & 3 & 4 & 0 & 1 \\ 0 & 0 & 2 & -2 & 0 \\ 0 & 0 & 0 & -1 & 3 \\ 0 & 0 & 0 & 0 & 5 \end{bmatrix} z = \begin{bmatrix} 1 \\ 1 \\ 0 \\ 2 \\ 3 \end{bmatrix}$$

by back substitution. The entries of the vector z are the entries of x^* corresponding to the pivot columns, so after substitution we have

$$x^* = \begin{bmatrix} z_1 \\ z_2 \\ z_3 \\ 0 \\ z_4 \\ z_5 \\ 0 \end{bmatrix} = \begin{bmatrix} 4/5 \\ 2/5 \\ -1/5 \\ 0 \\ -1/5 \\ 3/5 \\ 0 \end{bmatrix}.$$

To find all contributions from the null space, we compute a basis for the null space by repeatedly solving the homogeneous system $\mathsf{A}z = 0$. More precisely, in turn we set the entry of the vector x corresponding to one nonpivot column equal to 1 and the entry corresponding to all the other nonpivot column equal to zero, and we solve the homogeneous linear system for the remaining unknowns corresponding to the pivot columns. In the current example, where the nonpivot columns are the fourth and seventh columns, highlighted in boldface, and the equation that we need to solve for the unknown entries is

$$\begin{bmatrix} 1 & 3 & 3 & \mathbf{1} & 2 & 0 & \mathbf{0} \\ 0 & 3 & 4 & \mathbf{2} & 0 & 1 & \mathbf{1} \\ 0 & 0 & 2 & \mathbf{-1} & -2 & 0 & \mathbf{2} \\ 0 & 0 & 0 & \mathbf{0} & -1 & 3 & \mathbf{0} \\ 0 & 0 & 0 & \mathbf{0} & 0 & 5 & \mathbf{-2} \end{bmatrix} \begin{bmatrix} z_1 \\ z_1 \\ z_3 \\ \mathbf{1} \\ z_4 \\ z_5 \\ \mathbf{0} \end{bmatrix} = \begin{bmatrix} 0 \\ 0 \\ 0 \\ 0 \\ 0 \end{bmatrix}.$$

After moving the column vectors with known coefficients to the right-hand side, we find the values of the entries of z corresponding to the pivot columns by solving the linear system

$$\begin{bmatrix} 1 & 3 & 3 & 2 & 0 \\ 0 & 3 & 4 & 0 & 1 \\ 0 & 0 & 2 & -2 & 0 \\ 0 & 0 & 0 & -1 & 3 \\ 0 & 0 & 0 & 0 & 5 \end{bmatrix} z = \begin{bmatrix} -1 \\ -2 \\ 1 \\ 0 \\ 0 \end{bmatrix}$$

by back substitution and find the basis vector

$$z_1 = \begin{bmatrix} 3/2 \\ -4/3 \\ \mathbf{1} \\ 1/2 \\ 0 \\ 0 \\ \mathbf{0} \end{bmatrix}. \tag{10.17}$$

Likewise, to find the second basis vector of the null space, we set the seventh entry of z to one and the fourth to zero, and we solve

$$\begin{bmatrix} 1 & 3 & 3 & \mathbf{1} & 2 & 0 & \mathbf{0} \\ 0 & 3 & 4 & \mathbf{2} & 0 & 1 & \mathbf{1} \\ 0 & 0 & 2 & \mathbf{-1} & -2 & 0 & \mathbf{2} \\ 0 & 0 & 0 & \mathbf{0} & -1 & 3 & \mathbf{0} \\ 0 & 0 & 0 & \mathbf{0} & 0 & 5 & \mathbf{-2} \end{bmatrix} \begin{bmatrix} z_1 \\ z_2 \\ z_3 \\ \mathbf{0} \\ z_4 \\ z_5 \\ \mathbf{1} \end{bmatrix} = \begin{bmatrix} 0 \\ 0 \\ 0 \\ 0 \\ 0 \end{bmatrix}$$

for the unknown entries of z. Again, moving the vectors with known coefficients to the right-hand side, we have

$$\begin{bmatrix} 1 & 3 & 3 & 2 & 0 \\ 0 & 3 & 4 & 0 & 1 \\ 0 & 0 & 2 & -2 & 0 \\ 0 & 0 & 0 & -1 & 3 \\ 0 & 0 & 0 & 0 & 5 \end{bmatrix} z = \begin{bmatrix} 0 \\ -1 \\ -2 \\ 0 \\ 2 \end{bmatrix},$$

and after solving it we find the basis vector

$$z_2 = \begin{bmatrix} -4/5 \\ -11/15 \\ 1/5 \\ \mathbf{0} \\ 6/5 \\ 2/5 \\ \mathbf{1} \end{bmatrix}. \tag{10.18}$$

Once the particular solution x^* and the two basis vectors z_1 and z_2 of the null space have been computed, we have the general solution

$$x = \begin{bmatrix} 4/5 \\ 2/5 \\ -1/5 \\ 0 \\ -1/5 \\ 3/5 \\ 0 \end{bmatrix} + \alpha \begin{bmatrix} 3/2 \\ -4/3 \\ 1/2 \\ 1 \\ 0 \\ 0 \\ 0 \end{bmatrix} + \beta \begin{bmatrix} -4/5 \\ -11/15 \\ 1/5 \\ 0 \\ 6/5 \\ 2/5 \\ 1 \end{bmatrix}.$$

To summarize the results of this section, a solution of a linear system $\mathsf{A}x = b$ consisting of m equations in n unknowns exists if and only if, after the row operations applied to the matrix A and the vector b, the zero rows in the row reduced form of A correspond to zero entries in the transformed right-hand side vector.

If the linear system admits a solution and all columns are pivot columns, the solution is unique, and it can be computed by solving a lower triangular and an upper triangular linear system. If the linear system admits a solution and there are nonpivot columns, then there are infinitely many solutions, and the dimension of the null space of A equals the number of nonpivot columns. Each nonpivot column contributes one basis vector of the null space.

10.5 ▪ The four fundamental subspaces from a pivot perspective

The LU factorization makes it easy to identify linearly independent columns of the matrix A, as those correspond to the pivot columns of U. Since every vector in the range of A can be expressed as a linear combination of the pivot columns, the dimension of the range of A is equal to the number r of pivots, which is also the number of nonzero singular values.

Therefore, the number of pivots of a matrix A matches to the number of nonzero singular values, explaining why it is referred to as the *rank* of the matrix.

If a matrix A is invertible, the linear system $\mathsf{A}x = b$ has a unique solution for any choice of b. This implies that the general solution coincides with the particular solution, or, equivalently, that all columns are pivot columns. Moreover, the reduced form of the matrix A does not have any trailing rows of zeros; otherwise we could choose a vector b for which the solution does not exist. Therefore the matrix must have n nonzero singular values, and the dimension of the row space is $m = n$, implying that the matrix is square and invertible.

If we have the SVD of A at our disposal, the inverse of A is straightforward to compute, and the solution can be obtained by multiplying the vector b by A^{-1}.

If the linear system admits a solution and A has $n - r$ nonpivot columns, the matrix A has a nontrivial null space of dimension $n-r$. Once the matrix has been reduced to row echelon form, a basis of its null space can be computed by repeatedly solving the homogeneous linear system $\mathsf{A}x = 0$, each time setting all entries of x corresponding to nonpivot columns to zero except for one that is set to one.

If the SVD of the matrix A is available, the last $n - r$ columns of V are an orthonormal basis for $\mathcal{N}(\mathsf{A})$. This basis is orthonormal, while the basis computed from the row reduced form is, in general, not orthonormal.

The elimination process that takes the matrix A into its row reduced form is such that each nonzero row contains a pivot. Because of the upper triangular structure of the result and because of the fact that we are operating by taking linear combinations of the rows of A, the pivot rows of the reduced matrix, and hence those of the original one, are linearly independent. This implies that the dimension of the row space of A, or, equivalently, of the range of A^T, is equal to the number of pivots, and the solvability of the linear system for a specified vector b is the same as requesting that there be no vector $y \in \mathcal{N}\left(\mathsf{A}^\mathsf{T}\right)$ such that $y^\mathsf{T} b \neq 0$.

In conclusion, while the numerical values of the pivots and nonzero singular values of a matrix may be very different, their number is the same and is equal to the rank of the matrix. Unlike the calculation of the SVD of a matrix A, its reduction to upper echelon form can be carried out in a straightforward manner, and it provides a means to extract information about the four fundamental subspaces and to find bases for them.

10.6 ▪ Cholesky factorization

The general LU decomposition of a matrix A can carried out in a particular way for the special class of *symmetric positive definite* (SPD) matrices in $\mathbb{R}^{n\times n}$.

Definition 10.6. *A matrix* $\mathsf{A} \in \mathbb{R}^{n\times n}$ *is symmetric positive definite (SPD) if*

1. $\mathsf{A}^\mathsf{T} = \mathsf{A}$ *(symmetry),*
2. $v^\mathsf{T}\mathsf{A}v > 0$ *for every* $v \in \mathbb{R}^n$, $v \neq 0$ *(positive definiteness).*

As an example of an SPD matrix, assume that

$$\mathsf{A} = \mathsf{B}^\mathsf{T}\mathsf{B}, \quad \text{where } \mathsf{B} \in \mathbb{R}^{n\times n} \text{ is invertible.}$$

Then,

$$\mathsf{A}^\mathsf{T} = (\mathsf{B}^\mathsf{T}\mathsf{B})^\mathsf{T} = \mathsf{B}^\mathsf{T}\mathsf{B} = \mathsf{A},$$

and for any $v \neq 0$, we have

$$v^\mathsf{T}\mathsf{A}v = v^\mathsf{T}\mathsf{B}^\mathsf{T}\mathsf{B}v = (\mathsf{B}v)^\mathsf{T}(\mathsf{B}v) = \|\mathsf{B}v\|^2 > 0,$$

since the null space of an invertible matrix B cannot contain the nonvanishing vector v.

Every SPD matrix can be factorized in a symmetric way as

$$\mathsf{A} = \mathsf{R}^\mathsf{T}\mathsf{R}, \tag{10.19}$$

where $\mathsf{R} \in \mathbb{R}^{n\times n}$ is upper triangular with diagonal entries $r_{jj} > 0$. The decomposition is known as the *Cholesky factorization* of A, and the matrix R is the *Cholesky factor*.

The algorithm for determining the Cholesky factor R is rather straightforward. Assume that there is a real matrix R with positive diagonal entries such that

$$\mathsf{A} = \mathsf{R}^\mathsf{T}\mathsf{R} = \begin{bmatrix} r_{11} & & & \\ r_{12} & r_{22} & & \\ \vdots & & \ddots & \\ r_{1n} & r_{2n} & \cdots & r_{nn} \end{bmatrix} \begin{bmatrix} r_{11} & r_{12} & \cdots & r_{1n} \\ & r_{22} & \cdots & r_{2n} \\ & & \ddots & \vdots \\ & & & r_{nn} \end{bmatrix}.$$

In order for (10.19) to be satisfied for the first row, we must have

$$a_{11} = r_{11}^2, \quad \text{or } r_{11} = \sqrt{a_{11}},$$

and

$$r_{1j} = \frac{a_{1j}}{r_{11}}, \quad 2 \le j \le n.$$

This gives us a formula for determining the entries in the first row of R, the positive definiteness of A guaranteeing that $a_{11} > 0$. We then move to the second row of (10.19): because of the symmetry of the matrix A, it suffices to consider the components in its upper triangular part. Equating the diagonal components yields

$$a_{22} = r_{12}^2 + r_{22}^2, \quad \text{or} \quad r_{22} = \sqrt{a_{22} - r_{12}^2},$$

where r_{12} was already determined in the previous round and the positive definiteness of A guarantees that $a_{22} - r_{12}^2 > 0$, although this is not immediately obvious. The remaining entries of the second row of R must satisfy

$$r_{12}r_{1j} + r_{22}r_{2j} = a_{2j}, \quad 2 \le j \le n,$$

and hence

$$r_{2j} = \frac{a_{2j} - r_{12}r_{1j}}{r_{22}}.$$

The process continues in a similar manner for the subsequent rows until all entries of the matrix R have been computed.

A natural question arising at this point is how to check whether a symmetric matrix is positive definite. In general, there is no simple test for the positive definiteness, and a common practice is to try to compute the Cholesky decomposition! If the matrix is not positive definite, the process will stall because, at some point, the computation of the square root of a negative number will be attempted. Conversely, if the Cholesky decomposition is found, then for every $v \in \mathbb{R}^n$, $v \neq 0$,

$$v^\mathsf{T}\mathsf{A}v = v^\mathsf{T}\mathsf{R}^\mathsf{T}\mathsf{R}v = \|\mathsf{R}v\|^2 > 0,$$

since $\mathsf{R}v = 0$ is satisfied only for $v = 0$, because the diagonal entries of R are positive.

Problems

In Problems 1–7, the matrix A is

$$\mathsf{A} = \begin{bmatrix} 2 & -1 & 5 & 0 \\ 0 & 6 & 1 & -4 \\ 3 & 3 & 1 & 1 \\ 0 & 8 & 2 & 9 \\ -4 & 3 & 2 & 5 \end{bmatrix}.$$

1. Find a matrix D such the product of A and D reorders the rows A so that row 1 and row 3 are swapped, and so are row 2 and row 5. State how this is achieved by AD or by DA and verify that this is indeed the case by computing the matrix-matrix product.

2. Find a matrix D such the product of A and D reorders the columns A swapping column 1 and column 4, and column 2 and column 3. State if this is achieved by AD or by DA and verify that this is indeed the case by computing the matrix-matrix product.

3. Write the elementary matrix E_1 such that $\mathsf{E}_1\mathsf{A}$ is of the form

$$\mathsf{A}_1 = \mathsf{E}_1\mathsf{A} = \begin{bmatrix} * & * & * & * \\ 0 & * & * & * \\ 0 & * & * & * \\ 0 & * & * & * \\ 0 & * & * & * \end{bmatrix}$$

and compute A_1.

4. Find (by reasoning and subsequent checking) the inverse of E_1.

5. Write the elementary matrix E_2 such that $\mathsf{E}_2\mathsf{A}_1$ is of the form

$$\mathsf{A}_2 = \mathsf{E}_2\mathsf{E}_1\mathsf{A} = \begin{bmatrix} * & * & * & * \\ 0 & * & * & * \\ 0 & 0 & * & * \\ 0 & 0 & * & * \\ 0 & 0 & * & * \end{bmatrix}$$

and compute A_2.

6. Find (by reasoning and subsequent checking) the inverse of E_2.

7. Compute the matrix products
$\mathsf{E}_2\mathsf{E}_1$;
$\mathsf{E}_1^{-1}\mathsf{E}_2^{-1}$;
$\mathsf{E}_2^{-1}\mathsf{E}_1^{-1}$.

8. Let $\mathsf{A} = \mathsf{LU}$, where L and U are the lower and upper triangular matrices obtained by going through the process of Gaussian elimination. The upper triangular matrix U is the row reduced form of A. Show that

$$\mathcal{N}(\mathsf{A}) = \mathcal{N}(\mathsf{U}).$$

9. Show that if U is the row reduced form of the matrix A, then

$$\mathcal{R}(\mathsf{A}^\mathsf{T}) = \mathcal{R}(\mathsf{U}^\mathsf{T}).$$

10. Row reduce the matrix

$$\mathsf{A} = \begin{bmatrix} 2 & 1 & 0 & 5 & 1 & 2 & 2 \\ 4 & 3 & -1 & 2 & 2 & 6 & -4 \\ 0 & -1 & 1 & -3 & 0 & 1 & 3 \\ 0 & 3 & 1 & -1 & 0 & 3 & 2 \end{bmatrix}$$

and write the matrix representation of the elementary transformations that you apply.

(a) Find the rank and the dimension of the null space of the matrix A by reasoning on the row reduced matrix U.

(b) If r is the rank of the matrix U, identify the columns of the original matrix that span the range of A.

(c) Is the range of A the same as the range of U? Justify your answer.

(d) Find a basis for the row space and a basis for the null space of A.

(e) Find a basis for the row space and a basis for the null space of U.

11. Row reduce the matrix

$$\mathsf{A} = \begin{bmatrix} 3 & 2 & 1 \\ 9 & 8 & -6 \\ 3 & -5 & 0 \\ 0 & 2 & 6 \end{bmatrix},$$

and write the elementary transformations that you apply as matrices.

(a) Find the rank and the dimension of the null space of the matrix A by reasoning on the row reduced matrix U.

(b) If r is the rank of the matrix, identify the columns of the original matrix that span the range of A.

(c) Is the range of A the same as the range of U? Justify your answer.

(d) Find a basis for the row space and a basis for the null space of A.

(e) Find a basis for the row space and a basis for the null space of U.

Use the information above to decide whether the linear system

$$\mathsf{A}x = b, \quad b = \begin{bmatrix} 2 \\ -3 \\ 0 \\ 0 \end{bmatrix}$$

has a solution. If so, is the solution unique? Justify your answer.

12. Is the matrix

$$\begin{bmatrix} 1 & -2 & 0 \\ 0 & 1 & 3 \end{bmatrix}$$

row reduced? If not, complete the transformation, and then find

(a) the rank of the matrix;

(b) the dimension of its null space;

(c) a basis of the null space.

Is the matrix invertible? Justify your answer.

13. Is the matrix

$$\begin{bmatrix} 1 & 0 & 0 & 0 \\ 0 & 0 & 1 & -9 \\ 0 & 0 & 0 & 0 \end{bmatrix}$$

in row reduced echelon form (RREF)? If not, complete the transformation, and then find

(a) the rank of the matrix;

(b) the dimension of its null space;

(c) a basis of the null space.

14. Find the rank and the null space of the matrix

$$\mathsf{A} = \begin{bmatrix} 3 & 1 & 0 & 3 & 1 & -2 & 2 \\ 0 & 0 & 1 & 5 & -5 & -4 & 1 \\ 0 & 2 & -6 & 4 & 0 & 3 & 2 \\ 0 & 4 & 5 & 2 & 0 & 2 & 5 \end{bmatrix}$$

by row reducing it.

(a) Identify the columns of the original matrix that are a basis of the range.

(b) Find a basis for the null space of the original matrix.

(c) If $b = \begin{bmatrix} 2 \\ -1 \\ 3 \\ 0 \end{bmatrix}$, does the linear system $\mathsf{A}x = b$ have a solution? If so, find the general solution of the linear system. If not, explain why.

15. Find the general solution of the linear system $\mathsf{A}x = b$, where

$$\mathsf{A} = \begin{bmatrix} 1 & 1 & 6 & 3 \\ 3 & 0 & 4 & 1 \\ 0 & -5 & 7 & 4 \\ 1 & 3 & 5 & 0 \end{bmatrix}, \quad b = \begin{bmatrix} 5 \\ 2 \\ -1 \\ 4 \end{bmatrix}.$$

16. Given

$$\mathsf{A} = \begin{bmatrix} 1 & 1 & 6 & 3 \\ 3 & 0 & 4 & 1 \\ 0 & 4 & 10 & 1 \\ 1 & 2 & 5 & 0 \end{bmatrix}, \quad b = \begin{bmatrix} 27 \\ 8 \\ 5 \\ 23 \end{bmatrix},$$

find the general solution of the linear system $\mathsf{A}x = b$ if it exists; otherwise explain why it does not.

17. Let A be an $m \times n$ matrix with $m \geq n$ and linearly independent columns. Show that if $z_1, z_2, \ldots, z_k$ is a set of linearly independent vectors in $\mathbb{R}^n$, then $\mathsf{A}z_1, \mathsf{A}z_2, \ldots, \mathsf{A}z_k$ are linearly independent vectors in $\mathbb{R}^m$.

18. Given

$$\mathsf{A} = \begin{bmatrix} 1 & -1 & 0 & 2 & 3 \\ 2 & 0 & 2 & -1 & 0 \\ 4 & -2 & 2 & 3 & 6 \end{bmatrix}, \qquad \mathsf{b} = \begin{bmatrix} 1 \\ 1 \\ 5 \end{bmatrix},$$

find the general solution of the linear system $Ax = b$ if it exists; otherwise explain why it does not exist.

19. Given the matrix

$$\mathsf{A} = \begin{bmatrix} -1 & 2 & 1 & 0 & 2 \\ 2 & 0 & 0 & 3 & -1 \\ -1 & 6 & 3 & 3 & 5 \end{bmatrix}, \quad b = \begin{bmatrix} -1 \\ 0 \\ c \end{bmatrix}, \quad a \in \mathbb{R}.$$

   (a) For which values of c, if any, does the equation $\mathsf{A}x = b$ have a solution? Explain.

   (b) After choosing c so that the system has a solution, find a particular solution of the equation $\mathsf{A}x = b$.

20. For the matrix in A in the previous the problem, find

   (a) a basis for the row space;

   (b) a basis for the column space;

   (c) a basis for the null space.

21. Given the matrix factorization $\mathsf{A} = \mathsf{LU}$, where

$$\mathsf{A} = \begin{bmatrix} 4 & -3 & -1 & 5 & 2 \\ -16 & 12 & 2 & -17 & -7 \\ 8 & -6 & -12 & 22 & 10 \end{bmatrix},$$

and

$$\mathsf{L} = \begin{bmatrix} 1 & 0 & 0 \\ -4 & 1 & 0 \\ 2 & 5 & 1 \end{bmatrix}, \qquad \mathsf{U} = \begin{bmatrix} 4 & -3 & -1 & 5 & 2 \\ 0 & 0 & -2 & 3 & 1 \\ 0 & 0 & 0 & -3 & 1 \end{bmatrix}.$$

   (a) Determine the dimension of the column space of A and find a basis for it.

   (b) Determine the dimension of the null space of A and find a basis for it.

   (c) If $\mathsf{y} = \begin{bmatrix} 2 \\ 1 \\ -1 \end{bmatrix}$, find the general solution of $\mathsf{U}x = y$ if it exists; otherwise explain why it does not exist.

22. If

$$\mathsf{A} = \begin{bmatrix} 1 & 0 \\ 3 & 1 \end{bmatrix} \begin{bmatrix} 1 & -1 & 2 \\ 0 & -2 & -1 \end{bmatrix}, \quad b = \begin{bmatrix} 2 \\ 13 \end{bmatrix},$$

find the general solution to $\mathsf{A}x = b$.

23. Compute by hand the Cholesky factorization of the matrix

$$\mathsf{A} = \begin{bmatrix} 1 & 2 & 1 \\ 2 & 13 & 5 \\ 1 & 5 & 18 \end{bmatrix}.$$

Chapter 11

Determinants

The definition of the determinant requires that we delve into multilinear functions and is not immediately intuitive. In general, the direct calculation of determinants can be a very time consuming process that does not add much to our understanding of the matrix. That said, there are some important applications where determinants have a central role, thus justifying their presence even in a concise treatment of linear algebra.

We begin this chapter by formally defining multilinear functionals, which are mappings of several vectors in a vector space that are linear with respect to each vector. Classical examples of multilinear functionals are the real inner products and the determinant of a square matrix, where the input vectors are the columns of the matrix. The rest of the chapter is dedicated to presenting several useful results about determinants.

11.1 ▪ Linear functions and functionals

We start by recalling that a function

$$f : \mathbb{R}^n \to \mathbb{R}^m$$

is *linear* if for all $v, w \in \mathbb{R}^n$, $\alpha \in \mathbb{R}$,

$$f(\alpha v + w) = \alpha f(v) + f(w).$$

As already anticipated in Chapter 6, there is a natural connection between linear functions and matrices.

Theorem 11.1. *Every linear function f from $\mathbb{R}^n$ to $\mathbb{R}^m$ can be represented as an $m \times n$ matrix.*

Proof. Let $v_1, v_2, \ldots, v_n$ be a basis of $\mathbb{R}^n$, and $x = \sum_{j=1}^n x_j v_j$ the representation of an arbitrary vector $x \in \mathbb{R}^n$ in that basis. If

$$f(v_j) = a_j \in \mathbb{R}^m,$$

it follows from the linearity of the mapping f that

$$f(v) = f\left(\sum_{j=1}^n x_j v_j\right) = \sum_{j=1}^n x_j f(v_j) = \sum_{j=1}^n x_j a_j = \mathsf{A}x,$$

where

$$\mathsf{A} = \begin{bmatrix} \vdots & & \vdots \\ a_1 & \dots & a_n \\ \vdots & & \vdots \end{bmatrix}, \quad x = \begin{bmatrix} x_1 \\ \vdots \\ x_n \end{bmatrix}. \qquad \square$$

Now we are ready to extend the definition of linear functions to mappings from multiple copies of a vector space to the real or complex field $\mathbb{F}$.

Let V be a vector space over $\mathbb{F}$ of dimension n. A mapping

$$F : V^k \to \mathbb{F}$$

is a *k-linear form*, $1 \leq k \leq n$, if it is linear with respect to each component, that is, for $1 \leq j \leq k$,

$$F(v_1, \dots, \alpha v_j + w_j, \dots, v_k) = \alpha F(v_1, \dots, v_j, \dots, v_k) + F(v_1, \dots, w_j, \dots, v_k),$$

where $v_j \in V$, $\alpha \in \mathbb{F}$. Linear forms with $k = 1$ are called *linear functionals*; linear forms with $k = 2$ are called *bilinear forms*. An example of a bilinear form is the inner product in a real vector space.

With this definition in hand, we can prove the following result.

Theorem 11.2 (Riesz's Representation Theorem). *Every linear functional $F : V \to \mathbb{F}$ can be expressed as an inner product*

$$F(v) = q^{\mathsf{T}} x$$

for some vector $q \in \mathbb{F}^n$, where the entries of $x \in \mathbb{F}^n$ are the coordinates of the vector v in a given basis $\{v_j\}_{j=1}^n$. The vector $q \in \mathbb{F}^n$ is called the representation *of F in the basis $\{v_j\}_{j=1}^n$.*

Proof. Every $v \in V$ can be written as a linear combination of the basis vectors, $v = \sum_{j=1}^n x_j v_j$. Then it follows from the linearity of F that

$$F(v) = F\left(\sum_{j=1}^n x_j v_j\right) = \sum_{j=1}^n x_j F(v_j) = \sum_{j=1}^n x_j q_j = q^{\mathsf{T}} x,$$

where $q^{\mathsf{T}} = \begin{bmatrix} q_1 & \dots & q_n \end{bmatrix}$, $q_j = F(v_j)$. $\square$

In other words, the theorem says that it is always possible to represent a linear functional as an inner product with a fixed vector $q \in \mathbb{F}^n$. An analogous result holds for bilinear forms, as stated in the next theorem.

Theorem 11.3. *Every bilinear form $F : V \times V \to \mathbb{F}$ can be expressed in matrix-vector form as*

$$F(u, v) = x^{\mathsf{T}} \mathsf{B} y$$

for some matrix $\mathsf{B} \in \mathbb{F}^{n \times n}$. Here $x, y \in \mathbb{F}^n$ are the coordinate vectors of u and v in a basis $\{v_j\}_{j=1}^n$ of V. The matrix B is the representation of the bilinear form F in that basis.

Proof. Let $v_1, \ldots, v_n$ be a basis of V. Then

$$\begin{aligned} F(u,v) &= F\left(\sum_{j=1}^{n} x_j v_j, \sum_{\ell=1}^{n} y_\ell v_\ell\right) \\ &= \sum_{j=1}^{n}\sum_{\ell=1}^{n} x_j y_\ell F\left(v_j, v_\ell\right) \\ &= \sum_{j=1}^{n}\sum_{\ell=1}^{n} x_j y_\ell B_{j\ell} \\ &= x^{\mathsf{T}}\mathsf{B}y, \end{aligned}$$

where B is the $n \times n$ matrix with entries $\mathsf{B}_{j\ell} = F\left(v_j, v_\ell\right)$. □

A bilinear form is *symmetric* if swapping the two input vectors does not change its value, i.e.,

$$F(u,v) = F(v,u) \text{ for all } u, v \in V.$$

A bilinear form is *antisymmetric*, or *skew-symmetric*, if

$$F(u,v) = -F(v,u) \text{ for all } u, v \in V.$$

If B is the matrix representation of a bilinear form F, the bilinear form is symmetric if and only if B is a symmetric matrix,

$$\mathsf{B}^{\mathsf{T}} = \mathsf{B}.$$

Likewise, F is antisymmetric if and only if its representation is a skew-symmetric matrix,

$$\mathsf{B}^{\mathsf{T}} = -\mathsf{B}.$$

We remark that the diagonal entries of skew-symmetric matrices must vanish.

A multilinear form $F : V^k \to \mathbb{F}$ is *alternating* if it is antisymmetric with respect to swapping of any two vectors, that is, for any index pair (ℓ, j),

$$F(v_1, \ldots, v_j, \ldots, v_\ell, \ldots, v_k) = -F(v_1, \ldots, v_\ell, \ldots, v_j, \ldots, v_k).$$

For $k = 2$, alternating forms are antisymmetric forms.

11.2 ▪ Determinants

In this section, we consider the vector space $V = \mathbb{F}^n$, where $\mathbb{F} = \mathbb{R}$ or $\mathbb{F} = \mathbb{C}$.

Definition 11.4. *A* determinant *is an n-linear alternating form,*

$$F : \underbrace{V \times V \times \cdots \times V}_{n} \to \mathbb{F},$$

where $V = \mathbb{F}^n$, such that for the canonical basis vectors $e_1, \ldots, e_n$,

$$F(e_1, e_2, \ldots, e_n) = 1.$$

If $\mathsf{A} \in \mathbb{F}^{n\times n}$ is a matrix with column vectors $a_1, \ldots, a_n \in V = \mathbb{F}^n$, then

$$\det(\mathsf{A}) = F(a_1, \ldots, a_n).$$

This definition implicitly establishes the following three properties of the determinant.

1. If I_n is the $n \times n$ identity matrix, $\det(\mathsf{I}_m) = F(e_1, e_2, \ldots, e_n) = 1$.
2. It follows from the multilinearity of determinants that

$$\begin{aligned}&\det(\begin{bmatrix} a_1 & \cdots & \alpha a_j + b_j & \cdots & a_n \end{bmatrix})\\ &\quad = \alpha\det(\begin{bmatrix} a_1 & \cdots & a_j & \cdots & a_n \end{bmatrix}) + \det(\begin{bmatrix} a_1 & \cdots & b_j & \cdots & a_n \end{bmatrix}).\end{aligned}$$

3. The alternating property of determinants implies that for all index pairs (ℓ, j),

$$\begin{aligned}&\det(\begin{bmatrix} v_1 & \cdots & v_\ell & \cdots & v_j & \cdots & v_n \end{bmatrix})\\ &\quad = -\det(\begin{bmatrix} v_1 & \cdots & v_j & \cdots & v_\ell & \cdots & v_n \end{bmatrix}).\end{aligned}$$

All other properties of determinants can be derived from these three. We start with two basic results.

Theorem 11.5.

(i) If $\mathsf{A} \in \mathbb{F}^{n\times n}$ *is a diagonal matrix,*

$$\mathsf{A} = \begin{bmatrix} \alpha_1 & & \\ & \ddots & \\ & & \alpha_n \end{bmatrix},$$

then

$$\det(\mathsf{A}) = \alpha_1\alpha_2\cdots\alpha_n.$$

(ii) If $\mathsf{A} \in \mathbb{F}^{n\times n}$ *has two identical columns, then*

$$\det(\mathsf{A}) = 0.$$

Proof. Since the columns of a diagonal matrix A are scalar multiples of the canonical basis vectors, i.e., $a_j = \alpha_j e_j$ for $1 \leq j \leq n$, then

$$\begin{aligned}\det(\mathsf{A}) &= F(\alpha_1 e_1, \alpha_2 e_2, \ldots, \alpha_n e_n)\\ &= \alpha_1 F(e_1, \alpha_2 e_2, \ldots, \alpha_n e_n)\\ &= \alpha_1\alpha_2 F(e_1, e_2, \ldots, \alpha_n e_n)\\ &= \alpha_1\alpha_2\cdots\alpha_n F(e_1, e_2, \ldots, e_n)\\ &= \alpha_1\alpha_2\cdots\alpha_n,\end{aligned}$$

proving part (i) of the theorem.

To prove part (ii), assume for simplicity that $a_1 = a_2 = a$. Then, from the antisymmetry with respect to the first two positions it follows that

$$\begin{aligned}\det(\mathsf{A}) &= F(a, a, a_3, \ldots, a_n)\\ &= F(a_1, a_2, a_3, \ldots, a_n)\\ &= -F(a_2, a_1, a_3, \ldots, a_n)\\ &= -F(a, a, a_3, \ldots, a_n)\\ &= -\det(\mathsf{A}),\end{aligned}$$

and we conclude that

$$\det(\mathsf{A}) = 0,$$

as claimed. □

The following theorem generalizes part (ii) of the previous one.

Theorem 11.6. *If the columns of the matrix* $\mathsf{A} \in \mathbb{F}^{n\times n}$ *are not linearly independent, then*

$$\det(\mathsf{A}) = 0.$$

Proof. If the columns of A are not linearly independent, it is possible to express one of them as a linear combination of the others. For example, if

$$a_1 = \sum_{j=2}^{n} \mu_j a_j,$$

then

$$\det(\mathsf{A}) = F(\sum_{j=2}^{n} \mu_j a_j, a_2, \ldots, a_n) = \sum_{j=2}^{n} \mu_j F(a_j, a_2, \ldots, a_n) = 0,$$

because by Theorem 11.5, the determinant of a matrix with two equal columns vanishes. The same argument works if we replace any column of A by a linear combination of the others. □

The following corollary is an immediate consequence of Theorem 11.6.

Corollary 11.7. *If the determinant of* $\mathsf{A} \in \mathbb{F}^{n\times n}$ *is different from zero, then* A *is invertible.*

Proof. We prove the statement by contradiction. If a matrix A is not invertible, its columns are linearly dependent; hence from Theorem 11.6, $\det(\mathsf{A}) = 0$, contradicting the assumption. Therefore $\det(\mathsf{A}) \neq 0$ implies that A is invertible. □

The next result expresses the determinant of a matrix in terms of its entries. While in general this result is not of much use for computing of the determinant of matrices of dimension larger than three, the resulting formula is very useful to prove properties of the determinant of certain classes of matrices.

Expressing the first column $a_1 \in \mathbb{F}^n$ of A as a linear combination of the standard basis vectors,

$$a_1 = \begin{bmatrix} a_{1,1} \\ a_{2,1} \\ \vdots \\ a_{n,1} \end{bmatrix} = \sum_{k_1=1}^{n} a_{k_1,1} e_{k_1},$$

it follows from the multilinearity of the determinant that

$$\det(\mathsf{A}) = F\left(\sum_{k_1=1}^{n} a_{k_1,1} e_{k_1}, a_2, \ldots, a_n\right) = \sum_{k_1=1}^{n} a_{k_1,1} F(e_{k_1}, a_2, \ldots, a_n).$$

Repeating the process for the second column yields

$$\det(\mathsf{A}) = \sum_{k_1=1}^{n}\sum_{k_2=1}^{n} a_{k_1,1} a_{k_2,2} F(e_{k_1}, e_{k_2}, \ldots, a_n),$$

and proceeding inductively on the column index we conclude that

$$\det(\mathsf{A}) = \sum_{k_1=1}^{n}\sum_{k_2=1}^{n}\cdots\sum_{k_n=1}^{n} a_{k_1,1}a_{k_2,2}\cdots a_{k_n,n}F(e_{k_1},e_{k_2},\ldots,e_{k_n}). \tag{11.1}$$

Moreover, since if any two of the indices k_j are equal,

$$F(e_{k_1},e_{k_2},\ldots,e_{k_n}) = 0,$$

only the terms where all indices are different contribute to the sum.

Let $\mathscr{P}_n$ be the set of all $n!$ permutations of the indices $(1,2,\ldots,n)$. Denote a permutation $\sigma \in \mathscr{P}_n$ by

$$\sigma = (\sigma(1),\sigma(2),\ldots,\sigma(n)), \quad \sigma(j) \in \{1,2,\ldots,n\}, \quad \sigma(j) \neq \sigma(k), \text{ if } j \neq k.$$

We say that $\sigma \in \mathscr{P}_n$ is an *elementary permutation* if it swaps two indices, while keeping all other indices unchanged. It can be shown that every permutation can be expressed as a sequence of an even or odd number of elementary permutations. While the number of elementary permutations in different representations may vary, if it is even, or odd, for one representation, it is even or odd for every representation. In light of this observation, we say that a permutation $\sigma \in \mathscr{P}_n$ is *even* if it can be obtained by a sequence of an even number of elementary permutations, and it is *odd* otherwise.

It follows from the alternating property of the determinant that

$$\det(e_{\sigma(1)},e_{\sigma(2)},\ldots,e_{\sigma(n)}) = \begin{cases} 1 & \text{if } \sigma \text{ is an even permutation,} \\ -1 & \text{if } \sigma \text{ is an odd permutation.} \end{cases}$$

Introducing a function from $\mathscr{P}_n$ to $\{-1,1\}$, called the *sign* of a permutation, defined as

$$\operatorname{sign}(\sigma) = \begin{cases} 1 & \text{if } \sigma \text{ is an even permutation,} \\ -1 & \text{if } \sigma \text{ is an odd permutation,} \end{cases}$$

we obtain from (11.1) the following representation of the determinant.

Theorem 11.8. *The determinant of a matrix* $\mathsf{A} \in \mathbb{R}^{n\times n}$ *can be expressed as*

$$\det(\mathsf{A}) = \sum_{\sigma\in\mathscr{P}_n} \operatorname{sign}(\sigma) a_{\sigma(1),1}a_{\sigma(2),2}\cdots a_{\sigma(n),n},$$

where $\mathscr{P}_n$ *is the set of all permutations of the indices* $1,2,\ldots,n$.

Observe that each term of the sum is a product where every factor belongs to exactly one column of the matrix, and to the row whose index is the image in the permutation σ of the column index; each product is multiplied by the sign of the permutation.

The expression in the statement of the theorem coincides with the formula to compute the determinant of a 2×2 or 3×3 matrix found in many elementary textbooks. For $n = 2$, the formula reduces to

$$\det(\mathsf{A}) = a_{11}a_{22} - a_{12}a_{21}.$$

For $n = 3$, observe first that there are $3! = 6$ different permutations on the indices $1,2,3$,

$$\mathscr{P}_3 = \{(1,2,3),(1,3,2),(2,1,3),(2,3,1),(3,1,2),(3,2,1)\},$$

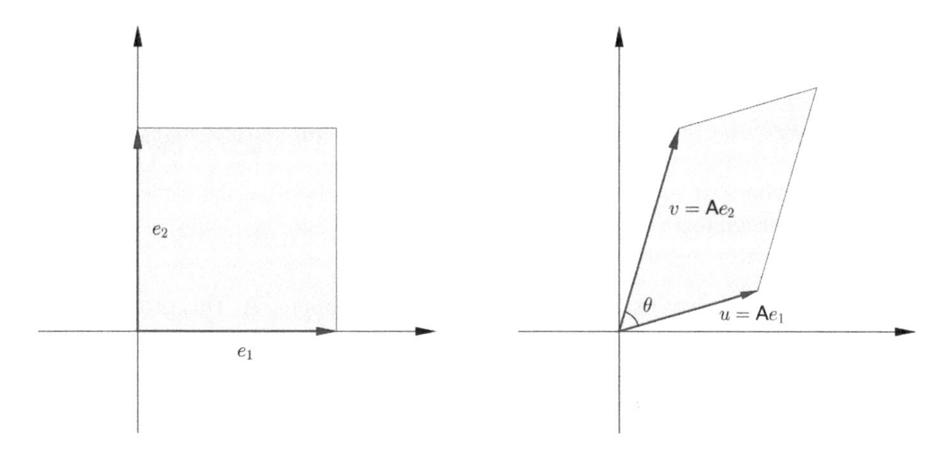

Figure 11.1. *A unit square spanned by canonical coordinate vectors (left) is mapped to a parallelogram (right) with area equal to* $|\det(\mathsf{A})|$.

whose signs are, respectively,

$$\{1, -1, -1, 1, 1, -1\}.$$

It follows from Theorem 11.8 that the determinant of a 3×3 matrix is

$$\det(\mathsf{A}) = a_{11}a_{22}a_{33} - a_{11}a_{32}a_{23} - a_{21}a_{12}a_{33} + a_{21}a_{32}a_{13} + a_{31}a_{12}a_{23} - a_{31}a_{22}a_{13}.$$

The computation of the determinant of a 3×3 matrix can be done by writing the 3×5 array obtained by adding the first two columns after the third one. The determinant of A is the sum of the products of the entries along the main diagonals, minus the products of the entries along the three antidiagonals; see the schematics below.

$$\begin{bmatrix} a_{11} & a_{12} & a_{13} \\ a_{21} & a_{22} & a_{23} \\ a_{31} & a_{32} & a_{33} \end{bmatrix} \begin{matrix} a_{11} & a_{12} \\ a_{21} & a_{22} \\ a_{31} & a_{32} \end{matrix}$$

Unfortunately such a shortcut does not generalize to matrices of larger dimensions.

The above formulas provide a geometric interpretation of the determinant. Consider first a unit square in $\mathbb{R}^2$, spanned by the canonical coordinate vectors e_1 and e_2. The image of this square under the mapping A is a parallelogram spanned by vectors $u = \mathsf{A}e_1$ and $v = \mathsf{A}e_2$. The area of the parallelogram, by elementary geometry, is given by

$$A = \|u\| \|v\| \sin\theta,$$

where θ is the angle between the vectors u and v; see Figure 11.1. On the other hand, the above formula coincides with the length of the cross product of u and v, and observing that u and v coincide with the columns of the matrix A, we have

$$A = \|u \times v\| = |u_1 v_2 - u_2 v_1| = |\det(\mathsf{A})|,$$

implying that the determinant of the matrix A represents the areal scaling of the space $\mathbb{R}^2$ under the linear map defined by A.

Similarly, in $\mathbb{R}^3$, the images of the canonical coordinate vectors, $u = \mathsf{A}e_1$, $v = \mathsf{A}e_2$, and $w = \mathsf{A}e_3$ span a parallelepiped whose volume is obtained by the triple product

$$V = |(u \times v) \cdot w| = |(u \times v)_1 w_1 + (u \times v)_2 w_2 + (u \times v)_3 v_3|.$$

Again, by observing that u, v, and w are the column vectors of A, the triple product formula leads to the conclusion that

$$V = |\det(\mathsf{A})|;$$

that is, the determinant gives the volume scaling under the mapping A. This geometric interpretation generalizes to all spatial dimensions, making the determinant extremely useful in geometry and measure theory, where the determinant of the Jacobian of a differentiable map represents the local volume scaling induced by the map.

The following corollary shows that the determinant of a matrix is equal to the determinant of its transpose.

Corollary 11.9. *For any $n \times n$ matrix* A

$$\det(\mathsf{A}) = \det(\mathsf{A}^\mathsf{T}).$$

Proof. It follows from Theorem 11.8 that

$$\det(\mathsf{A}^\mathsf{T}) = \sum_{\sigma \in \mathscr{P}_n} \operatorname{sign}(\sigma) a^\mathsf{T}_{\sigma(1),1} a^\mathsf{T}_{\sigma(2),2} \cdots a^\mathsf{T}_{\sigma(n),n},$$

where we use the notation

$$a^\mathsf{T}_{k,j} = (\mathsf{A}^\mathsf{T})_{k,j} = a_{j,k}.$$

Since the inverse of a permutation $\sigma \in \mathscr{P}_n$ is also a permutation $\tau \in \mathscr{P}_n$ such that $\tau(\sigma(j)) = j$, and $\operatorname{sign}(\tau) = \operatorname{sign}(\sigma)$, then

$$\begin{aligned}
\det(\mathsf{A}^\mathsf{T}) &= \sum_{\sigma \in \mathscr{P}_n} \operatorname{sign}(\sigma) a^\mathsf{T}_{\sigma(1),1} a^\mathsf{T}_{\sigma(2),2} \cdots a^\mathsf{T}_{\sigma(n),n} \\
&= \sum_{\sigma \in \mathscr{P}_n} \operatorname{sign}(\sigma) a_{1,\sigma(1)} a_{2,\sigma(2)} \cdots a_{n,\sigma(n)} \\
&= \sum_{\tau \in \mathscr{P}_n} \operatorname{sign}(\tau) a_{\tau(1),1} a_{\tau(2),2} \cdots a_{\tau(n),n} \\
&= \det(\mathsf{A}),
\end{aligned}$$

proving the result. □

The next result is also a direct consequence of Theorem 11.8.

Corollary 11.10. *If* $\mathsf{A} \in \mathbb{R}^{n \times n}$ *is an upper or lower triangular matrix, then*

$$\det(\mathsf{A}) = a_{11} a_{22} \cdots a_{nn}.$$

Proof. If the matrix A is upper triangular, there is only one way to select a nonzero element from the first column, namely the first one, and from that it follows that the only way to choose a nonzero element from the second column not in the first row is to pick the element on the diagonal. Proceeding inductively in this manner we show that the determinant of an upper triangular matrix contains only one term, namely the product of its diagonal entries. The proof that the determinant of a lower triangular matrix is the product of the diagonal entries follows from the observation that a lower triangular matrix is the transpose of an upper triangular matrix and an application of Corollary 11.9. ☐

Next we prove that the determinant of the product of two matrices is the product of the determinants of the factors.

Theorem 11.11. *If* $\mathsf{A}, \mathsf{B} \in \mathbb{F}^{n\times n}$*, then*

$$\det(\mathsf{AB}) = \det(\mathsf{A})\det(\mathsf{B}).$$

Proof. If we express the matrix B in terms of its columns,

$$\mathsf{B} = \begin{bmatrix} b_1 & b_2 & \cdots & b_n \end{bmatrix},$$

the matrix-matrix product can be written in terms of matrix-vector products as

$$\mathsf{AB} = \begin{bmatrix} \mathsf{A}b_1 & \mathsf{A}b_2 & \cdots & \mathsf{A}b_n \end{bmatrix},$$

and therefore,

$$\det(\mathsf{AB}) = F(\mathsf{A}b_1, \mathsf{A}b_2, \ldots, \mathsf{A}b_n).$$

On the other hand, each matrix-vector product $\mathsf{A}b_j$ is a linear combination of the columns a_k of the matrix A, that is,

$$\mathsf{A}b_j = \sum_{k_j=1}^{n} b_{k_j,j} a_{k_j}, \quad j = 1, \ldots, n.$$

Substituting this expression for $\mathsf{A}b_j$ in the determinant and using the multilinearity of the determinant, we have

$$\det(\mathsf{AB}) = \sum_{k_1=1}^{n} \sum_{k_2=1}^{n} \cdots \sum_{k_n=1}^{n} b_{k_1,1} b_{k_2,2} \cdots b_{k_n,n} F(a_{k_1}, a_{k_2}, \ldots, a_{k_n}).$$

Each term of the sum where a column of the matrix A is repeated vanishes; thus the only nonvanishing terms are those corresponding to permutations of the indices $1, 2, \ldots, n$; hence

$$\det(\mathsf{AB}) = \sum_{\sigma\in\mathscr{P}_n} b_{\sigma(1),1} b_{\sigma(2),2} \cdots b_{\sigma(n),n} F(a_{\sigma(1)}, a_{\sigma(2)}, \ldots, a_{\sigma(n)}).$$

Given a permutation σ, we have

$$F(a_{\sigma(1)}, a_{\sigma(2)}, \ldots, a_{\sigma(n)}) = \operatorname{sign}(\sigma) F(a_1, a_2, \ldots, a_n) = \operatorname{sign}(\sigma)\det(\mathsf{A});$$

therefore

$$\det(\mathsf{AB}) = \det(\mathsf{A}) \sum_{\sigma\in\mathscr{P}_n} \operatorname{sign}(\sigma) b_{\sigma(1),1} b_{\sigma(2),2} \cdots b_{\sigma(n),n} = \det(\mathsf{A})\det(\mathsf{B}),$$

completing the proof. ☐

The following three corollaries of Theorem 11.11 are very useful to compute determinants without having to resort to Theorem 11.8.

Corollary 11.12. *If* $\mathsf{A} = \mathsf{LU}$ *is the LU factorization of the* $n \times n$ *matrix* A*, then*

$$\det(\mathsf{A}) = \det(\mathsf{U}) = u_{1,1} \ldots u_{n,n}.$$

Proof. The result follows from the observation that L and U are triangular matrices, with the diagonal entries of L equal to 1, so that the determinant of L is also 1. □

Corollary 11.13. *If* A *is an invertible matrix,*

$$\det(\mathsf{A}^{-1}) = \frac{1}{\det(\mathsf{A})}.$$

Proof. This is an immediate consequence of the fact that $\mathsf{A}\mathsf{A}^{-1} = \mathsf{I}$ and that $\det(\mathsf{I}) = 1$. From Theorem 11.11,

$$\det(\mathsf{A})\det\left(\mathsf{A}^{-1}\right) = \det\left(\mathsf{A}\mathsf{A}^{-1}\right) = 1 \quad \Rightarrow \quad \det\left(\mathsf{A}^{-1}\right) = \frac{1}{\det(\mathsf{A})}. \qquad \square$$

Corollary 11.14. *The determinant of an orthogonal matrix is* ± 1.

Proof. If $\mathsf{U} \in \mathbb{R}^n$ is an orthogonal matrix, it holds that

$$1 = \det\left(\mathsf{I}_n\right) = \det\left(\mathsf{U}\mathsf{U}^\mathsf{T}\right) = \det\left(\mathsf{U}\right)\det\left(\mathsf{U}^\mathsf{T}\right) = \left(\det\left(\mathsf{U}\right)\right)^2,$$

from which the claim follows. □

The above reasoning gives us an alternative proof for Corollary 11.7.

Corollary 11.15. *If* A *is an invertible matrix, then* $\det(\mathsf{A}) \neq 0$.

Proof. If $\det(\mathsf{A}) = 0$ and A is invertible, then

$$1 = \det(\mathsf{A}\mathsf{A}^{-1}) = \det(\mathsf{A})\det(\mathsf{A}^{-1}) = 0,$$

leading to a contradiction. Therefore the determinant of A cannot vanish, which completes the proof. □

Problems

1. Let

$$\mathsf{A} = \begin{bmatrix} a & 1 & 0 \\ 0 & 1 & c \\ 0 & 0 & d \end{bmatrix},$$

   where $a,\ b,\ c$ are arbitrary constants.

   (a) What is the determinant of A? Justify your answer.

   (b) For which values of $a,\ b,\ c$ is A invertible? What is the determinant of A^{-1}?

2. If A is a 7×7 matrix, find $\det(-2\mathsf{A})$. Justify your answer.

3. If $\mathsf{A} = \mathsf{U}\Sigma\mathsf{V}^\mathsf{T}$ is the singular value decomposition of $\mathsf{A} \in \mathbb{R}^{n\times n}$, show that the determinant of A is the product of its singular values.

4. Two matrices A, B are similar if there exists an invertible matrix S such that

$$\mathsf{A} = \mathsf{SBS}^{-1}.$$

   Show that $\det(\mathsf{A}) = \det(\mathsf{B})$.

5. Show that if P is an $n \times n$ permutation matrix, then its determinant equals the sign of the corresponding permutation σ.

6. Show that if A is a symmetric positive definite matrix, its determinant is positive.

7. Show that if $\det(\mathsf{A}) = 0$, then the matrix A has a nontrivial null space.

8. Show that if A is an $n \times n$ matrix and $\mathsf{PA} = \mathsf{LU}$ is the LU factorization of A with the rows permuted according to P, then $\det(\mathsf{A}) = \pm\det(\mathsf{U})$.

9. Let $\mathsf{A} = \mathsf{QR}$ be the QR factorization of the real $n \times n$ matrix A.

   (a) What can you say about the determinant of Q?

   (b) Express the determinant of A in terms of the determinant of R.

   (c) Express the determinant of AA^T in terms of the determinant of R.

   (d) If $\det(\mathsf{R}) = 3$, is A invertible? If so, what is the determinant of A^{-1}?

Chapter 12

Eigenvalues and Eigenvectors

Eigenvectors and eigenvalues are a central topic in linear algebra. In fact, in addition to their theoretical importance, eigenvectors and eigenvalues play a key role in many applications, including dynamical systems, data science, and modern physics.

An eigenvector v of a matrix A is a vector whose direction remains unchanged upon multiplication by the matrix A. Eigenvectors and eigenvalues are defined only for square matrices because the only way $\mathsf{A}v$ could be a scalar multiple of v is if the two vectors have the same number of components, which is the same as requiring the matrix A to be square. Thus, the only difference between an eigenvector v and $\mathsf{A}v$ is a scaling by a complex number λ, referred to the eigenvalue associated with the eigenvector v.

In general, eigenvectors and eigenvalues are complex even if the underlying matrix has all real entries. In this chapter we derive general properties of eigenvectors and eigenvalues, and we characterize matrices whose eigenvectors and eigenvalues satisfy certain properties.

12.1 ▪ Definitions and basic properties

An $n \times n$ matrix $\mathsf{A} \in \mathbb{R}^{n\times n}$ or $\mathbb{C}^{n\times n}$ can be thought of as a mapping of vectors from $\mathbb{C}^n$ to vectors in $\mathbb{C}^n$. In this case the domain and the range of the linear function defined by A are subspaces of the same vector space $\mathbb{C}^n$. In this chapter, we are mostly focusing on real matrices but consider their action on complex vectors.

Definition 12.1. *A vector $v \in \mathbb{C}^n$, $v \neq 0$, is an* eigenvector *of $\mathsf{A} \in \mathbb{R}^{n\times n}$ or $\mathsf{A} \in \mathbb{C}^{n\times n}$ if it identifies a direction that is invariant under multiplication by A. In other words, if v is an eigenvector of A, there exists a scalar $\lambda \in \mathbb{C}$ such that*

$$\mathsf{A}v = \lambda v.$$

The scalar λ is the eigenvalue of A associated with the eigenvector v, and the pair (λ, v) is an eigenpair of A.

We start with the following basic result.

Theorem 12.2. *A scalar $\lambda \in \mathbb{C}$ is an eigenvalue and a vector $v \in \mathbb{C}^n$, $v \neq 0$, is an associated eigenvector of the matrix $\mathsf{A} \in \mathbb{R}^{n\times n}$ or $\mathsf{A} \in \mathbb{C}^{n\times n}$ if and only if*

$$(\mathsf{A} - \lambda \mathsf{I}_n)v = 0.$$

The number of linearly independent eigenvectors of A *associated with the eigenvalue* λ *is the dimension of the null space of* $\mathsf{A} - \lambda\mathsf{I}_n$.

Proof. If λ is an eigenvalue of A and $v \neq 0$ is a corresponding eigenvector, then

$$\mathsf{A}v - \lambda\mathsf{I}_n v = (\mathsf{A} - \lambda\mathsf{I}_n)\, v = 0 \Leftrightarrow v \in \mathcal{N}(\mathsf{A} - \lambda\mathsf{I}_n).$$

A basis of $\mathcal{N}(\mathsf{A} - \lambda\mathsf{I}_n)$ can be computed according to the following procedure:

1. Row reduce the matrix $\mathsf{A} - \lambda\mathsf{I}_n$ to upper echelon form by elementary row operations, and identify all nonpivot columns. Since $\mathsf{A} - \lambda\mathsf{I}_n$ is singular, there is at least one nonpivot column.

2. Solve the linear system $(\mathsf{A}-\lambda\mathsf{I}_n)z = 0$, setting all the entries corresponding to all nonpivot columns to zero, except for one that is set to one. This is a basis vector of the null space of $\mathsf{A} - \lambda\mathsf{I}_n$.

3. Repeat Step 2, changing the nonpivot columns corresponding to a nonzero entry in the solution. The number of vectors determined in such a manner is equal to the number of nonpivot columns.

The vectors determined in this manner are linearly independent, and each one is an eigenvector of A associated with the eigenvalue λ. □

Corollary 12.3. *Any nonzero scalar multiple of an eigenvector* v *of* A *associated with the eigenvalue* λ *is also an eigenvector of* A *associated with the eigenvalue* λ.

Proof. The geometrical intuition behind this result is that since an eigenvector identifies a direction that remains unchanged under multiplication by the matrix A, this property is independent of the length of the vector. Therefore, once we have found an eigenvector, we can scale it as we please. Formally, for any scalar $\beta \neq 0$,

$$\mathsf{A}v = v \Rightarrow \mathsf{A}(\beta v) = \beta\mathsf{A}v = \beta\lambda v = \lambda(\beta v),$$

thus proving that βv is an eigenvector of A associated with the eigenvalue λ. □

Corollary 12.4. *The set of eigenvectors of a square matrix* A *associated with the eigenvalue* λ *is a subspace, referred to as the* eigenspace *of* λ.

Proof. If v_1, v_2 are eigenvectors of A associated with the same eigenvalue λ, then

$$\mathsf{A}v_1 = \lambda v_1, \quad \mathsf{A}v_2 = \lambda v_2.$$

For any $\alpha, \beta \in \mathbb{C}$,

$$\begin{aligned}\mathsf{A}\,(\alpha v_1 + \beta v_2) &= \alpha\mathsf{A}v_1 + \beta\mathsf{A}v_2\\ &= \alpha\lambda v_1 + \beta\lambda v_2\\ &= \lambda\,(\alpha v_1 + \beta v_2)\,;\end{aligned}$$

therefore the subset of eigenvectors associated with the eigenvalue λ is closed under linear combinations. □

The eigenvalues of a matrix can be thought of as scalar representatives of the matrix itself when limited to the eigenspace. A number of operations on the matrix translate into the corresponding operations on the eigenvalues, such as computing the inverse or a power of a matrix. Some of these basic properties are stated in the following theorems.

Theorem 12.5. *If a square matrix* A *is invertible, all its eigenvalues are different from zero.*

Proof. We prove the statement by contradiction. Assume that A is invertible. If zero is an eigenvalue of the matrix A, and $v \neq 0$ is an associated eigenvector, then $\mathsf{A}v = 0$. From the invertibility of A, it follows that

$$v = \mathsf{A}^{-1}\mathsf{A}v = 0,$$

contradicting the assumption that $v \neq 0$. This proves that the statement must hold. □

The following theorem is closely related to the previous one.

Theorem 12.6. *If* A *is an invertible matrix and* λ *is one of its eigenvalues with associated eigenvector* v*, then* λ^{-1} *is an eigenvalue of* A^{-1} *and* v *is an associated eigenvector.*

Proof. It follows from $\mathsf{A}v = \lambda v$ and the invertibility of A that

$$v = \mathsf{A}^{-1}\mathsf{A}v = \lambda\mathsf{A}^{-1}v,$$

and further, from $\lambda \neq 0$ that

$$\mathsf{A}^{-1}v = \lambda^{-1}v,$$

thus proving the claim. □

Theorem 12.6 establishes a correspondence between the inverse of an invertible matrix and the reciprocal of its eigenvalues. The next theorem relates the powers of a matrix to the powers of its eigenvalues.

Theorem 12.7. *If* (λ, v) *is an eigenpair of a matrix* A*, then* (λ^k, v)*,* $k \geq 1$*, is an eigenpair of the matrix* A^k*.*

Proof. The proof is by induction on k. For $k = 1$ the statement is true by the definition of eigenpair.

Assume that the statement holds for $k - 1$ for some $k > 1$, that is,

$$\mathsf{A}^{k-1}v = \lambda^{k-1}v.$$

It follows from the associativity of matrix multiplication and the induction hypothesis that

$$\mathsf{A}^k v = \mathsf{A}\left(\mathsf{A}^{k-1}v\right) = \mathsf{A}\lambda^{k-1}v = \lambda^{k-1}\mathsf{A}v = \lambda^k v,$$

thus completing the induction. □

This theorem has an immediate generalization.

Corollary 12.8. *Let* p *be a polynomial of degree* k*,* $p(t) = \sum_{j=0}^{k} a_j t^j$*, and let* (λ, v) *be an eigenpair of the matrix* A*. Then* $(p(\lambda), v)$ *is an eigenpair of the matrix*

$$p(\mathsf{A}) = a_0\mathsf{I}_n + a_1\mathsf{A} + a_2\mathsf{A}^2 + \cdots + a_k\mathsf{A}^k.$$

Proof. The proof follows from a direct application of Theorem 12.7, since

$$\begin{aligned} p(\mathsf{A})v &= a_0 v + a_1 \mathsf{A} v + a_2 \mathsf{A}^2 v + \cdots + a_k \mathsf{A}^k v \\ &= a_0 v + a_1 \lambda v + a_2 \lambda^2 v + \cdots + a_k \lambda^k v \\ &= p(\lambda) v, \end{aligned}$$

as claimed. ◻

There are classes of matrices whose eigenvalues and eigenvectors satisfy special properties. We begin by considering orthogonal and unitary matrices.

Theorem 12.9. *If $\lambda \in \mathbb{C}$ is an eigenvalue of an orthogonal or unitary matrix, then $|\lambda| = 1$.*

Proof. The length preserving property of orthogonal and unitary matrices implies that if $v \neq 0$ is an eigenvector of an orthogonal or unitary matrix Q,

$$\mathsf{Q} v = \lambda v,$$

we have

$$\|v\|_2 = \|\mathsf{Q} v\|_2 = |\lambda| \|v\|_2,$$

and since $\|v\|_2 \neq 0$, we must have

$$|\lambda| = 1.$$

This completes the proof. ◻

The following theorem establishes an important connection between the singular values and singular vectors of a matrix and the eigenpairs of two square matrices associated with it.

Theorem 12.10. *Let A be an $m \times n$ real matrix with singular value decomposition*

$$\mathsf{A} = \mathsf{U} \Sigma \mathsf{V}^\mathsf{T},$$

and singular values $\sigma_1 \geq \sigma_2 \geq \cdots \geq \sigma_r \geq 0$, $r = \min\{m, n\}$. Then

1. *the scalars $\sigma_1^2, \sigma_2^2, \ldots, \sigma_r^2$ are eigenvalues of $\mathsf{A}\mathsf{A}^\mathsf{T}$ and $\mathsf{A}^\mathsf{T}\mathsf{A}$;*
2. *the right singular vector $v_j \in \mathbb{R}^n$, $j = 1, \ldots, r$, is an eigenvector of $\mathsf{A}^\mathsf{T}\mathsf{A} \in \mathbb{R}^{n \times n}$ associated with the eigenvalue σ_j^2;*
3. *the left singular vector $u_j \in \mathbb{R}^m$, $j = 1, \ldots, r$, is an eigenvector of $\mathsf{A}\mathsf{A}^\mathsf{T} \in \mathbb{R}^{m \times m}$ associated with the eigenvalue σ_j^2.*

Proof. We start by writing

$$\mathsf{A}^\mathsf{T}\mathsf{A} = \mathsf{V} \Sigma^\mathsf{T} \Sigma \mathsf{V}^\mathsf{T}, \quad \mathsf{A}\mathsf{A}^\mathsf{T} = \mathsf{U} \Sigma \Sigma^\mathsf{T} \mathsf{U}^\mathsf{T},$$

where $\Sigma^\mathsf{T}\Sigma$ is the $n \times n$ matrix,

$$\Sigma^\mathsf{T}\Sigma = \operatorname{diag}\left(\sigma_1^2, \ldots, \sigma_n^2\right) \quad \text{if } n \leq m,$$

or

$$\Sigma^\mathsf{T}\Sigma = \operatorname{diag}\left(\sigma_1^2, \ldots, \sigma_m^2, 0, \ldots 0\right) \quad \text{if } n > m.$$

Similarly, $\Sigma\Sigma^\mathsf{T}$ is the $m \times m$ matrix,

$$\Sigma\Sigma^\mathsf{T} = \operatorname{diag}\left(\sigma_1^2, \ldots, \sigma_n^2, 0, \ldots 0\right) \quad \text{if } n < m,$$

or

$$\Sigma\Sigma^\mathsf{T} = \operatorname{diag}\left(\sigma_1^2, \ldots, \sigma_m^2\right) \quad \text{if } n \geq m.$$

From the expressions above and the orthogonality of the columns of V it follows that

$$\begin{aligned}
\mathsf{A}^\mathsf{T}\mathsf{A}v_j &= \mathsf{V}\begin{bmatrix} \sigma_1^2 & & & & \\ & \ddots & & & \\ & & \sigma_r^2 & & \\ & & & \ddots & \\ & & & & 0 \end{bmatrix}_{n\times n} \mathsf{V}^\mathsf{T} v_j \\
&= \mathsf{V}\begin{bmatrix} \sigma_1^2 & & & & \\ & \ddots & & & \\ & & \sigma_r^2 & & \\ & & & \ddots & \\ & & & & 0 \end{bmatrix} e_j \\
&= \begin{cases} \sigma_j^2 v_j & \text{if } j \leq r, \\ 0 & \text{otherwise,} \end{cases}
\end{aligned}$$

proving the claim for $\mathsf{A}^\mathsf{T}\mathsf{A}$. The proof for $\mathsf{A}\mathsf{A}^\mathsf{T}$ is similar. □

We remark that the above connection between the SVD and eigenpairs is often used as a tool to derive the SVD.

12.2 ▪ Eigenvalues and roots of polynomials

If $v \neq 0$ is an eigenvector corresponding to an eigenvalue λ of the $n \times n$ matrix A, then, by Theorem 12.2, $v \in \mathcal{N}(\mathsf{A} - \lambda\mathsf{I}_n) = \mathcal{N}(\lambda\mathsf{I}_n - \mathsf{A})$. This implies that the null space of $\lambda\mathsf{I}_n - \mathsf{A}$ is nontrivial; hence $\det(\lambda\mathsf{I}_n - \mathsf{A}) = 0$. Therefore the eigenvalues of A are the zeros of the polynomial

$$p_\mathsf{A}(\lambda) = \det(\lambda\mathsf{I}_n - \mathsf{A}),$$

called the *characteristic polynomial* of the matrix A.

The relation between the eigenvalues of a matrix and the zeros of its characteristic polynomial could lead us to believe that a way to compute the eigenvalues of A would be to find the roots of its characteristic polynomial. This approach relies on the assumption that there is a straightforward way to compute the roots of polynomials, which is only the case for very small values of n. In fact, the celebrated Abel–Ruffini theorem states that there is no closed form solution for a general polynomial equation of degree five or higher, making this powerful connection between the eigenvalues of a matrix and the roots of its characteristic polynomial more of theoretical than of algorithmic interest.

The characteristic polynomial of a matrix $\mathsf{A} \in \mathbb{C}^{n\times n}$ is of degree n, and according to the Fundamental Theorem of Algebra, it can be factorized as the product of n monomials

$$p_\mathsf{A}(\lambda) = (\lambda - \lambda_1)(\lambda - \lambda_2)\cdots(\lambda - \lambda_n),$$

where the scalars $\lambda_1, \ldots, \lambda_n \in \mathbb{C}$ are the n roots of the characteristic polynomial, and hence the eigenvalues of the matrix A. If $\mathsf{A} \in \mathbb{R}^{n \times n}$, the coefficients of the characteristic polynomial are real numbers; therefore the eigenvalues of A either are real or must appear in complex conjugate pairs. However, in general there is no guarantee that they are real roots.

Some of the roots of the characteristic polynomial may coincide. The number of times the monomial $(\lambda - \lambda_j)$ appears in the factorization is called the multiplicity of the root λ_j.

Definition 12.11. *For a matrix* $\mathsf{A} \in \mathbb{C}^{n \times n}$,

1. *the* algebraic multiplicity *of the eigenvalue* λ_j*, denoted by* $m_a(\lambda_j, \mathsf{A})$*, is the multiplicity of* λ *as a root of the characteristic polynomial;*

2. *the* geometric multiplicity *of the eigenvalue* $\lambda_j \in \mathbb{C}$*, denoted by* $m_g(\lambda_j, \mathsf{A})$*, is the dimension of the null space* $\mathcal{N}(\lambda_j \mathsf{I}_n - \mathsf{A})$ *or, equivalently, the maximum number of linearly independent associated eigenvectors.*

An important relation between algebraic and geometric multiplicity is established in the following theorem.

Theorem 12.12. *The algebraic multiplicity of an eigenvalue cannot be less than the geometric multiplicity; that is, for all eigenvalues* $\mu \in \mathbb{C}$,

$$m_a(\mu, \mathsf{A}) \geq m_g(\mu, A).$$

Proof. Let $m_g(\mu, \mathsf{A}) = k \leq n$, and let $v_1, \ldots, v_k \in \mathbb{C}^n$ be an orthonormal basis spanning the eigenspace,

$$H_\mu = \{v \in \mathbb{C}^n \mid \mathsf{A}v = \mu v\} = \mathcal{N}(\mu \mathsf{I}_n - \mathsf{A}).$$

Using Theorem 8.1, we complete the set of vectors into an orthonormal basis $\{v_1, \ldots, v_n\}$ of $\mathbb{C}^n$. Denoting by V the unitary matrix

$$\mathsf{V} = \begin{bmatrix} v_1 & \cdots & v_n \end{bmatrix},$$

we have, for $\lambda \in \mathbb{C}$,

$$\begin{aligned} &\mathsf{V}^{\mathsf{H}}(\lambda \mathsf{I}_n - \mathsf{A})\mathsf{V} \\ &= \begin{bmatrix} v_1^{\mathsf{H}} \\ \vdots \\ v_n^{\mathsf{H}} \end{bmatrix} \begin{bmatrix} (\lambda - \mu)v_1 & \cdots & (\lambda - \mu)v_k & (\lambda \mathsf{I}_n - \mathsf{A})v_{k+1} & \cdots & (\lambda \mathsf{I}_n - \mathsf{A})v_n \end{bmatrix} \\ &= \begin{bmatrix} (\lambda - \mu)\mathsf{I}_k & \mathsf{B}_1(\lambda) \\ \mathsf{O}_{n \times (n-k)} & \mathsf{B}_2(\lambda) \end{bmatrix}, \end{aligned}$$

where

$$\begin{bmatrix} \mathsf{B}_1(\lambda) \\ \mathsf{B}_2(\lambda) \end{bmatrix} = \mathsf{V}^{\mathsf{H}}(\lambda \mathsf{I}_n - \mathsf{A}) \begin{bmatrix} v_{k+1} & \cdots & v_n \end{bmatrix}.$$

The determinant of the left side of the equation yields

$$\begin{aligned} \det(\mathsf{V}^{\mathsf{H}}(\lambda \mathsf{I}_n - \mathsf{A})\mathsf{V}) &= \det(\mathsf{V}^{\mathsf{H}})\det(\lambda \mathsf{I}_n - \mathsf{A})\det(\mathsf{V}) \\ &= \det(\mathsf{V}^{\mathsf{H}}\mathsf{V})\det(\lambda \mathsf{I}_n - \mathsf{A}) \\ &= \det(\lambda \mathsf{I}_n - \mathsf{A}), \end{aligned}$$

and by equating it to the determinant of the right side, we obtain

$$\det(\lambda \mathsf{I}_n - \mathsf{A}) = (\lambda - \mu)^k \det(\mathsf{B}_2(\lambda)),$$

proving that μ is a root of at least multiplicity k of the characteristic polynomial; thus $m_a(\mu, \mathsf{A}) \geq k$. □

To see that in general, it is possible that $m_a(\mathsf{A}, \mu) > m_g(\mathsf{A}, \mu)$, consider the matrix

$$\mathsf{A} = \begin{bmatrix} \mu & 1 \\ 0 & \mu \end{bmatrix}.$$

Clearly,

$$\det(\lambda \mathsf{I}_2 - \mathsf{A}) = (\lambda - \mu)^2,$$

showing that $m_a(\mathsf{A}, \mu) = 2$. To find the geometric multiplicity, we consider the equation

$$\mathsf{A}v = \mu v \Leftrightarrow \begin{cases} \mu v_1 + v_2 = \mu v_1 \\ \mu v_2 = \mu v_2 \end{cases} \Leftrightarrow v = \begin{bmatrix} v_1 \\ 0 \end{bmatrix},$$

implying that the eigenspace is one-dimensional.

If

$$m_a(\lambda, \mathsf{A}) > m_g(\lambda, A),$$

we say that the eigenvalue λ is *defective*. A matrix is defective if at least one of its eigenvalues is defective: only matrices with repeated eigenvalues can be defective. A complete analysis of the eigenvalues and eigenvectors of defective matrices would require the introduction of the Jordan canonical form, a topic that we are not addressing in this book.

Another important connection between matrices and polynomials is established by the following well-known theorem.

Theorem 12.13 (Cayley–Hamilton). *Let* A *be an* $n \times n$ *matrix and* $p_{\mathsf{A}}(t)$ *its characteristic polynomial. Then* $p_{\mathsf{A}}(\mathsf{A}) = 0$.

The observation that the eigenvalues of a matrix are the roots of its characteristic polynomial provides a path from matrices to polynomials. The reverse journey, from polynomials to matrices, translates the problem of finding the zeros of a polynomial into the problem of computing the eigenvalues of a matrix. More specifically, given the monic polynomial of degree n,

$$p_n(\lambda) = a_0 + a_1\lambda + \cdots + a_{n-1}\lambda^{n-1} + \lambda^n,$$

we define its *companion matrix*,

$$\mathsf{C} = \begin{bmatrix} 0 & & & & & -a_o \\ 1 & 0 & & & & -a_1 \\ 0 & 1 & 0 & & & \\ & & & & & \\ & & & & 0 & -a_{n-2} \\ 0 & \dots & & 0 & 1 & -a_{n-1} \end{bmatrix}.$$

It is possible to show that

$$\det(\lambda \mathsf{I}_n - \mathsf{C}) = a_0 + a_1\lambda + \cdots + a_{n-1}\lambda^{n-1} + \lambda^n;$$

therefore the polynomial $p_n(\lambda)$ is the characteristic polynomial of the matrix C. In practice, this means that finding the roots of a polynomial can always be recast as an eigenvalue problem, and vice versa. This connection allows the implementation of effective root-finding algorithms using linear algebraic methods.

We conclude this section with an example showing how to determine the geometric multiplicity of repeated eigenvalues.

The matrices

$$\mathsf{A}_1 = \begin{bmatrix} 2 & 0 & 0 & 0 \\ 0 & 2 & 0 & 0 \\ 0 & 0 & 4 & 0 \\ 0 & 0 & 0 & 3 \end{bmatrix}, \quad \mathsf{A}_2 = \begin{bmatrix} 2 & 1 & 0 & 0 \\ 0 & 2 & 0 & 0 \\ 0 & 0 & 4 & 0 \\ 0 & 0 & 0 & 3 \end{bmatrix}$$

have the same eigenvalues $2, 2, 4, 3$. In both cases the eigenvectors corresponding to the eigenvalues 3 and 4, which are scalar multiples of the canonical basis vectors e_4 and e_3, respectively, are mutually orthogonal and hence linearly independent. The matrix A_1 has two linearly independent eigenvectors corresponding to the eigenvalue $\lambda = 2$, since the first two columns of the matrix

$$2\,\mathsf{I}_4 - \mathsf{A}_1 = \begin{bmatrix} 0 & 0 & 0 & 0 \\ 0 & 0 & 0 & 0 \\ 0 & 0 & -2 & 0 \\ 0 & 0 & 0 & -1 \end{bmatrix}$$

are nonpivot columns, implying that the two linearly independent vectors $x_1 = e_1$ and $x_2 = e_2$ are a basis of its null space. Therefore the repeated eigenvalue $\lambda = 2$ has algebraic and geometric multiplicity 2, and the matrix A_1 is nondefective.

The eigenvectors of the matrix A_2 corresponding to the eigenvalue $\lambda = 2$ are in the null space of

$$2\,\mathsf{I}_4 - \mathsf{A}_2 = \begin{bmatrix} 0 & -1 & 0 & 0 \\ 0 & 0 & 0 & 0 \\ 0 & 0 & -2 & 0 \\ 0 & 0 & 0 & -1 \end{bmatrix},$$

which consists of all scalar multiples of the vector $x_1 = e_1$. In this case the dimension of $\mathcal{N}(2\,\mathsf{I}_4 - \mathsf{A}_2)$ is one, hence the geometric multiplicity of the eigenvalue $\lambda = 2$ is one, smaller than its algebraic multiplicity, and the matrix A_2 is defective.

12.3 ▪ Eigenvalues of symmetric and Hermitian matrices

In general, the eigenvalues and eigenvectors of a matrix A are complex numbers, even if the entries of the matrix are all real. This can be easily verified by computing the eigenvalues and eigenvectors of the matrix

$$\mathsf{A} = \begin{bmatrix} 0 & -1 \\ 1 & 0 \end{bmatrix}.$$

If λ is an eigenvalue of A and $v \neq 0$ is an associated eigenvector,

$$\begin{bmatrix} 0 & -1 \\ 1 & 0 \end{bmatrix} \begin{bmatrix} v_1 \\ v_2 \end{bmatrix} = \lambda \begin{bmatrix} v_1 \\ v_2 \end{bmatrix} \Rightarrow \begin{bmatrix} -\lambda & -1 \\ 1 & -\lambda \end{bmatrix} \begin{bmatrix} v_1 \\ v_2 \end{bmatrix} = \begin{bmatrix} 0 \\ 0 \end{bmatrix}.$$

The characteristic polynomial of A is

$$\det(\lambda \mathsf{I} - \mathsf{A}) = \det\left(\begin{bmatrix} \lambda & 1 \\ -1 & \lambda \end{bmatrix} \right) = \lambda^2 + 1 = 0,$$

and its eigenvalues are $\lambda = \pm i$.

In this section we identify some classes of matrices whose eigenvalues and eigenvectors have some special properties. We begin with a class of matrices with real eigenvalues.

Lemma 12.14. *For any matrix* $\mathsf{A} \in \mathbb{C}^{n\times n}$ *and* $x, y \in \mathbb{C}^n$,

$$\langle \mathsf{A}x, y\rangle = \langle x, \mathsf{A}^{\mathsf{H}} y\rangle.$$

Proof. Recalling that $\mathsf{A}^{\mathsf{H}} = \overline{\mathsf{A}}^{\mathsf{T}}$ it follows from the definition of natural inner product in $\mathbb{C}^n$ that

$$\langle \mathsf{A}x, y\rangle = x^{\mathsf{T}}\mathsf{A}^{\mathsf{T}}\overline{y} = x^{\mathsf{T}}\overline{\mathsf{A}}^{\mathsf{H}}\overline{y} = x^{\mathsf{T}}\overline{\mathsf{A}^{\mathsf{H}}y} = \langle x, \mathsf{A}^{\mathsf{H}}y\rangle,$$

as claimed. □

Theorem 12.15.

(i) If A *is an* $n \times n$ *real symmetric matrix, or a complex Hermitian matrix, then its eigenvalues are all real.*

(ii) For a real symmetric matrix, it is possible to find real associated eigenvectors.

Proof. If λ is an eigenvalue of A and x is an associated eigenvector, then $\mathsf{A}x = \lambda x$. If $\langle\,\cdot\,,\,\cdot\,\rangle$ is the standard inner product in $\mathbb{C}^n$,

$$\langle \mathsf{A}x, x\rangle = \langle \lambda x, x\rangle = \lambda\|x\|_2^2,$$

and from Lemma 12.14,

$$\langle \mathsf{A}x, x\rangle = \langle x, \mathsf{A}^{\mathsf{H}}x\rangle = \langle x, \mathsf{A}x\rangle = \overline{\lambda}\|x\|_2^2,$$

implying that

$$\lambda\|x\|^2 = \overline{\lambda}\|x\|^2, \text{ and hence } \lambda = \overline{\lambda},$$

because $x \neq 0$. This proves that the eigenvalues of a symmetric matrix, or, more generally, a Hermitian matrix, are real.

To prove that if a real matrix has a real eigenvalue, then there is a corresponding real eigenvector, let $x \in \mathbb{C}^n$ be an eigenvector associated with the real eigenvalue λ. Then

$$x = x_{\mathrm{Re}} + ix_{\mathrm{Im}},$$

where $x_{\mathrm{Re}} \in \mathbb{R}^n$ is the vector of the real parts of the entries of x, and $x_{\mathrm{Im}} \in \mathbb{R}^n$ is the vector of the imaginary components, and

$$\mathsf{A}x = \mathsf{A}x_{\mathrm{Re}} + i\mathsf{A}x_{\mathrm{Im}} = \lambda(x_{\mathrm{Re}} + ix_{\mathrm{Im}}) = \lambda x_{\mathrm{Re}} + i\lambda x_{\mathrm{Im}}.$$

Equating real and imaginary parts on both sides and recalling that A, λ, x_{Re}, and x_{Im} are real, we have

$$\mathsf{A}x_{\mathrm{Re}} = \lambda x_{\mathrm{Re}} \text{ and } \mathsf{A}x_{\mathrm{Im}} = \lambda x_{\mathrm{Im}}.$$

Therefore, if $x_{\mathrm{Re}} \neq 0$, it is also a real eigenvector of A associated with the eigenvalue λ. On the other hand, if $x_{\mathrm{Re}} = 0$, then $x_{\mathrm{Im}} \neq 0$ is the sought real eigenvector, thus completing the proof. □

An alternative, constructive proof of the existence of real eigenvectors is based on the observation that if x is an eigenvector of A associated with the eigenvalue λ, then

$$\mathsf{A}x = \lambda x \quad \Rightarrow \quad x \in \mathcal{N}(\lambda\mathsf{I}_n - \mathsf{A}).$$

The computation of a basis vector of the null space of the real matrix $\lambda \mathsf{I}_n - \mathsf{A}$ outlined in chapter 10 begins by reducing the matrix to upper echelon form via row reduction operations. This guarantees that the reduced matrix is real. The next step consists of setting the entries of the vector corresponding to nonpivot columns to zero or one and solving the associated homogeneous linear system to find the values of the entries corresponding to the pivot columns. The calculations involve only the four basic operations of real numbers, and therefore the basis vectors of the null space of $\mathsf{A} - \lambda \mathsf{I}_n$ are necessarily real.

The fact that real symmetric matrices have real eigenvalues and eigenvectors makes it easier to interpret these quantities in geometric terms. The following theorem establishes an important geometric relation between pairs of eigenvectors corresponding to distinct eigenvalues of real symmetric matrices.

Theorem 12.16. *Let $\mathsf{A} \in \mathbb{R}^{n\times n}$ be a symmetric matrix, and let (λ, u) and (μ, v) be two eigenpairs with $u, v \in \mathbb{R}^n$ and $\lambda \neq \mu$. Then the vectors u and v are mutually orthogonal.*

Proof. It follows from Theorem 12.15 that since A is real and symmetric, $\lambda, \mu \in \mathbb{R}$. Multiplying both sides of the identity

$$\mathsf{A}v = \mu v$$

from the left by u^{T} yields

$$u^{\mathsf{T}}\mathsf{A}v = \mu u^{\mathsf{T}}v, \tag{12.1}$$

and using the symmetry of A, we have that

$$u^{\mathsf{T}}\mathsf{A}v = u^{\mathsf{T}}\mathsf{A}^{\mathsf{T}}v = (\mathsf{A}u)^{\mathsf{T}}v = \lambda u^{\mathsf{T}}v. \tag{12.2}$$

Therefore, setting the right-hand sides of (12.1) and (12.2) equal, we obtain

$$(\lambda - \mu)u^{\mathsf{T}}v = 0,$$

which, in light of the fact that $\lambda \neq \mu$, implies that $u^{\mathsf{T}}v = 0$, or, equivalently, that u and v are mutually orthogonal. □

It turns out that an even stronger result concerning the eigenvectors of real symmetric matrices, or more generally, Hermitian matrices, can be obtained, namely that the eigenvectors can be defined so that they form an orthonormal basis. We give two different proofs of this important result, one based on a matrix factorization and another based on properties of the eigenspaces.

To motivate the discussion, let A be an $n \times n$ matrix with eigenvalues $\lambda_1, \ldots, \lambda_n$, and assume that the eigenvalues are associated with a set of n linearly independent eigenvectors $v_1, \ldots, v_n$. Introduce the matrix V, whose columns are the eigenvectors of A, and the diagonal matrix Λ, whose diagonal entries are the corresponding eigenvalues of A. Then,

$$\begin{aligned} \mathsf{AV} &= \mathsf{A}\begin{bmatrix} v_1 & \cdots & v_n \end{bmatrix} \\ &= \begin{bmatrix} \mathsf{A}v_1 & \cdots & \mathsf{A}v_n \end{bmatrix} \\ &= \begin{bmatrix} \lambda_1 v_1 & \cdots & \lambda_n v_n \end{bmatrix} \\ &= \mathsf{V}\Lambda, \end{aligned}$$

where

$$\Lambda = \operatorname{diag}(\lambda_1, \ldots, \lambda_n) \in \mathbb{C}^{n\times n},$$

and since V is invertible,

$$\mathsf{A} = \mathsf{V}\Lambda\mathsf{V}^{-1}. \tag{12.3}$$

The decomposition (12.3), if it exists, is called the *spectral factorization* or *eigenvalue decomposition* of the matrix A. However, not all matrices admit a spectral factorization, and those that do are called *diagonalizable*, the term referring to the transformation of the matrix to a diagonal form.

We recall that two $n \times n$ matrices A and B are *similar* if there exists an invertible matrix S, called the *similarity transformation*, such that

$$\mathsf{SAS}^{-1} = \mathsf{B}. \tag{12.4}$$

The eigenvalue decomposition, if it exists, is therefore a similarity transformation, and the matrix A is similar to a diagonal matrix. More generally, the following result holds for similar matrices.

Theorem 12.17. *If A and B are $n \times n$ similar matrices, then they have the same eigenvalues. Moreover, the similarity transformation defines a bijection between eigenvectors associated with the same eigenvalues.*

Proof. If A, B satisfy (12.4) and λ is an eigenvalue of A with eigenvector v, then $\mathsf{A}v = \lambda v$ and

$$\lambda \mathsf{S}v = \mathsf{S}\lambda v = \mathsf{SA}v = \mathsf{SAS}^{-1}\mathsf{S}v = \mathsf{BS}v;$$

therefore λ is an eigenvalue of B with corresponding eigenvector $\mathsf{S}v \neq 0$. □

Next we prove that every $n \times n$ complex matrix is similar to an upper triangular matrix via a unitary similarity matrix.

Lemma 12.18 (Schur's lemma). *For every complex $n \times n$ matrix A there is a unitary matrix U and an upper triangular matrix T such that*

$$\mathsf{U}^{\mathsf{H}}\mathsf{AU} = \mathsf{T}. \tag{12.5}$$

Proof. We proceed by induction. For $n = 1$ the factorization (12.5) exists trivially. Now assume that (12.5) holds for $(n-1) \times (n-1)$ matrices, and let A be an $n \times n$ matrix. Let λ be an eigenvalue of A, and let v be a corresponding eigenvector scaled to have unit norm. Complete v into an orthonormal basis of $\mathbb{C}^n$, and let $\widetilde{\mathsf{U}} = \begin{bmatrix} v & \mathsf{V} \end{bmatrix}$ be the unitary matrix whose columns are the basis vectors. Then

$$\widetilde{\mathsf{U}}^{\mathsf{H}}\mathsf{A}\widetilde{\mathsf{U}} = \begin{bmatrix} \lambda & v^{\mathsf{H}}\mathsf{AV} \\ 0 & \mathsf{V}^{\mathsf{H}}\mathsf{AV} \end{bmatrix}.$$

Since $\mathsf{V}^{\mathsf{H}}\mathsf{AV}$ is an $(n-1) \times (n-1)$ matrix, it follows from the induction hypothesis that there exist a unitary matrix W and an upper triangular matrix $\widetilde{\mathsf{T}}$ such that

$$\mathsf{V}^{\mathsf{H}}\mathsf{AV} = \mathsf{W}\widetilde{\mathsf{T}}\mathsf{W}^{\mathsf{H}}.$$

The matrix $\mathsf{U} = \begin{bmatrix} v & \mathsf{VW} \end{bmatrix}$ is unitary, since

$$\begin{aligned} \mathsf{U}^{\mathsf{H}}\mathsf{U} &= \begin{bmatrix} v^{\mathsf{H}} \\ \mathsf{W}^{\mathsf{H}}\mathsf{V}^{\mathsf{H}} \end{bmatrix} \begin{bmatrix} v & \mathsf{VW} \end{bmatrix} \\ &= \begin{bmatrix} v^{\mathsf{H}}v & v^{\mathsf{H}}\mathsf{VW} \\ \mathsf{W}^{\mathsf{H}}\mathsf{V}^{\mathsf{H}}v & \mathsf{W}^{\mathsf{H}}\mathsf{V}^{\mathsf{H}}\mathsf{VW} \end{bmatrix} \\ &= \begin{bmatrix} 1 & 0 \\ 0 & \mathsf{W}^{\mathsf{H}}\mathsf{W} \end{bmatrix} = \mathsf{I}_n, \end{aligned}$$

and

$$\begin{aligned}
\mathsf{U}^{\mathsf{H}}\mathsf{A}\mathsf{U} &= \begin{bmatrix} v^{\mathsf{H}} \\ \mathsf{W}^{\mathsf{H}}\mathsf{V}^{\mathsf{H}} \end{bmatrix} \mathsf{A} \begin{bmatrix} v & \mathsf{V}\mathsf{W} \end{bmatrix} \\
&= \begin{bmatrix} v^{\mathsf{H}} \\ \mathsf{W}^{\mathsf{H}}\mathsf{V}^{\mathsf{H}} \end{bmatrix} \begin{bmatrix} \lambda v & \mathsf{A}\mathsf{V}\mathsf{W} \end{bmatrix} \\
&= \begin{bmatrix} \lambda & v^{\mathsf{H}}\mathsf{A}\mathsf{V}\mathsf{W} \\ 0 & \mathsf{W}^{\mathsf{H}}(\mathsf{V}^{\mathsf{H}}\mathsf{A}\mathsf{V})\mathsf{W} \end{bmatrix} \\
&= \begin{bmatrix} \lambda & v^{\mathsf{H}}\mathsf{A}\mathsf{V}\mathsf{W} \\ 0 & \widetilde{\mathsf{T}} \end{bmatrix} = \mathsf{T},
\end{aligned}$$

completing the proof. □

Formula (12.5) is called the *Schur factorization.* Schur factorization implies immediately the following important result.

Theorem 12.19. *Every $n \times n$ Hermitian matrix A has real eigenvalues and a set of n orthogonal eigenvectors. In particular, every Hermitian matrix is diagonalizable by a unitary similarity transformation.*

Proof. It follows from the Schur factorization and the conjugate symmetry of A that

$$\mathsf{T} = \mathsf{U}^{\mathsf{H}}\mathsf{A}\mathsf{U} = \mathsf{U}^{\mathsf{H}}\mathsf{A}^{\mathsf{H}}\mathsf{U} = \mathsf{T}^{\mathsf{H}}. \tag{12.6}$$

We observe that the entries of T under the diagonal vanish, and likewise, the entries of T^{H} above the diagonal vanish. Therefore, the identity (12.6) implies that T must be diagonal. Moreover, by Theorem 12.15, we know that the eigenvalues of A are real, and by Theorem 12.17 they coincide with eigenvalues of T that further coincide with the diagonal entries of T, so T is real diagonal, hence proving the statement. □

The above proof based on Schur's lemma may not be very intuitive, and therefore, we give an alternative proof that relies on properties of projections and subspaces.

Alternative proof of Theorem 12.19. Let $\lambda_1, \lambda_2, \dots, \lambda_k$ denote the distinct real eigenvalues of a symmetric matrix $\mathsf{A} \in \mathbb{R}^{n\times n}$, $k \le n$, with algebraic multiplicities

$$m_a(\mathsf{A}, \lambda_j) = n_j;$$

that is, the characteristic polynomial of A is

$$\det(\lambda \mathsf{I}_n - \mathsf{A}) = (\lambda - \lambda_1)^{n_1} \cdots (\lambda - \lambda_k)^{n_k},$$

and $n_1 + \cdots + n_k = n$. We denote by H_j the eigenspace associated to eigenvalue λ_j,

$$H_j = \{v \in \mathbb{C}^n \mid \mathsf{A}v = \lambda_j v\}, \quad 1 \le j \le k.$$

The geometric multiplicity of the eigenvector λ_j is, by definition,

$$m_g(\mathsf{A}, \lambda_j) = \dim(H_j) = n_j' \le n_j.$$

From the orthogonality of eigenvectors associated to different eigenvectors of a Hermitian matrix, it follows that

$$H_j \perp H_\ell, \quad j \ne \ell.$$

In each H_j, we find n'_j mutually orthogonal eigenvectors associated to the eigenvalue λ_j. The collection of the eigenvectors associated with all eigenvalues is an orthogonal basis of the space

$$\mathscr{H} = H_1 \oplus \cdots \oplus H_k \subset \mathbb{R}^n.$$

We claim that this basis, in fact, is a basis of $\mathbb{R}^n$. To demonstrate this, consider the orthocomplement of $\mathscr{H}$,

$$\mathscr{H}^\perp = \{v \in \mathbb{C}^n \mid \langle v, w\rangle = 0 \text{ for all } w \in \mathscr{H}\}.$$

We note first that $\mathscr{H}^\perp$ is an invariant subspace of A,

$$\mathsf{A}(\mathscr{H}^\perp) \subset \mathscr{H}^\perp,$$

which is seen as follows: If $w \in \mathscr{H}^\perp$, then for every $v \in \mathscr{H}$,

$$\langle v, \mathsf{A}w\rangle = \langle \mathsf{A}v, w\rangle = 0,$$

because $\mathsf{A}v \in \mathscr{H}$, being a linear combination of eigenvectors of A. Therefore, $\mathsf{A}w \in \mathscr{H}^\perp$. Assume now that $\mathscr{H}^\perp \neq \{0\}$. Then, we find a set of orthonormal vectors that span the space,

$$\mathscr{H}^\perp = \text{span}\{w_1, \ldots, w_m\}.$$

We denote by $\mathsf{W} \in \mathbb{C}^{n\times m}$ the matrix with columns w_j,

$$\mathsf{W} = \begin{bmatrix} w_1 & \cdots & w_m \end{bmatrix},$$

and define

$$\mathsf{A}_\mathsf{W} = \mathsf{W}^\mathsf{H}\mathsf{A}\mathsf{W} \in \mathbb{C}^{m\times m}.$$

The matrix A_W is Hermitian, so it has real eigenvalues. Let μ be an eigenvalue of A_W and let $z \in \mathbb{C}^m$ denote an associated eigenvector,

$$\mathsf{A}_\mathsf{W} z = \mu z, \quad z \neq 0.$$

Left multiplication of both sides of this identity by W gives

$$\mathsf{W}\mathsf{A}_\mathsf{W} z = \mu(\mathsf{W}z). \tag{12.7}$$

We observe that since $w = \mathsf{W}z \in \mathscr{H}^\perp$, and this space is an invariant subspace, $\mathsf{A}w \in \mathscr{H}^\perp$. Moreover, since $\mathsf{W}\mathsf{W}^\mathsf{H}$ is the orthogonal projector on $\mathscr{H}^\perp$, then

$$\mathsf{W}\mathsf{A}_\mathsf{W} z = (\mathsf{W}\mathsf{W}^\mathsf{H}) \underbrace{\mathsf{A}(\mathsf{W}z)}_{=\mathsf{A}w \in \mathscr{H}^\perp} = \mathsf{A}w.$$

We conclude that by (12.7), therefore,

$$\mathsf{A}w = \mu w, \quad w \neq 0,$$

so μ must be one of the eigenvalues of A, say, $\mu = \lambda_1$. But in this case $w \in H_1$, which is in contradiction with $w \perp H_1$, thus completing the proof. $\square$

12.4 ▪ Normal matrices and unitary diagonalizability

In general there is no guarantee that a matrix has a full set of linearly independent eigenvectors, and even for diagonalizable matrices, eigenvectors are not necessarily mutually orthogonal. The class of matrices that admit a complete set of orthogonal eigenvectors is fully characterized by the following easy to check condition.

Definition 12.20. *A matrix* $\mathsf{A} \in \mathbb{C}^{n\times n}$ *is* normal *if it commutes with its conjugate transpose,*

$$\mathsf{A}\mathsf{A}^{\mathsf{H}} = \mathsf{A}^{\mathsf{H}}\mathsf{A}.$$

The following lemma plays an important role in characterizing matrices with a full set of orthogonal eigenvectors.

Lemma 12.21. *If an upper triangular matrix is normal, then it is diagonal.*

Proof. Let $\mathsf{T} \in \mathbb{C}^{n\times n}$ be an upper triangular matrix such that $\mathsf{T}\mathsf{T}^{\mathsf{H}} = \mathsf{T}^{\mathsf{H}}\mathsf{T}$. Then,

$$\begin{bmatrix} t_{11} & \cdots & t_{1n} \\ & \ddots & \vdots \\ & & t_{nn} \end{bmatrix} \begin{bmatrix} \overline{t_{11}} & & \\ \vdots & \ddots & \\ \overline{t_{1n}} & \cdots & \overline{t_{nn}} \end{bmatrix} = \begin{bmatrix} \overline{t_{11}} & & \\ \vdots & \ddots & \\ \overline{t_{1n}} & \cdots & \overline{t_{nn}} \end{bmatrix} \begin{bmatrix} t_{11} & \cdots & t_{1n} \\ & \ddots & \vdots \\ & & t_{nn} \end{bmatrix}$$

and, in particular,

$$\left(\mathsf{T}\mathsf{T}^{\mathsf{H}}\right)_{11} = \sum_{j=1}^{n} t_{1j}\overline{t_{1j}} = \sum_{j=1}^{n} |t_{1j}|^2 = \left(\mathsf{T}^{\mathsf{H}}\mathsf{T}\right)_{11} = |t_{11}|^2,$$

implying that

$$t_{1j} = 0, \ j = 2, \ldots, n.$$

In a similar manner we proceed to prove that $t_{k\ell} = 0$ for all k with $\ell > k$, thus completing the proof. □

Theorem 12.22. *A matrix* A *is unitarily diagonalizable if and only if* A *is normal.*

Proof. We first prove that unitarily diagonalizable matrices are normal. Let $\mathsf{A} = \mathsf{U}\Lambda\mathsf{U}^{\mathsf{H}}$, where Λ is a diagonal matrix. Then

$$\begin{aligned} \mathsf{A}\mathsf{A}^{\mathsf{H}} &= \mathsf{U}\Lambda\mathsf{U}^{\mathsf{H}}\mathsf{U}\Lambda^{\mathsf{H}}\mathsf{U}^{\mathsf{H}} = \mathsf{U}\Lambda\Lambda^{\mathsf{H}}\mathsf{U}^{\mathsf{H}} \\ &= \mathsf{U}\Lambda^{\mathsf{H}}\Lambda\mathsf{U}^{\mathsf{H}} = \mathsf{A}^{\mathsf{H}}\mathsf{A}. \end{aligned}$$

Conversely, the proof that a normal matrix is unitarily diagonalizable is based on Schur's lemma. Let

$$\mathsf{A} = \mathsf{U}\mathsf{T}\mathsf{U}^{\mathsf{H}},$$

where U is unitary and T is upper triangular. After replacing A by its Schur decomposition,

$$\mathsf{A}\mathsf{A}^{\mathsf{H}} = \mathsf{U}\mathsf{T}\mathsf{U}^{\mathsf{H}}\left(\mathsf{U}\mathsf{T}\mathsf{U}^{\mathsf{H}}\right)^{\mathsf{T}} = \mathsf{U}\mathsf{T}\mathsf{U}^{\mathsf{H}}\mathsf{U}\mathsf{T}^{\mathsf{H}}\mathsf{U}^{\mathsf{H}} = \mathsf{U}\mathsf{T}\mathsf{T}^{\mathsf{H}}\mathsf{U}^{\mathsf{H}},$$

and

$$\mathsf{A}^{\mathsf{H}}\mathsf{A} = (\mathsf{UTU}^{\mathsf{H}})^{\mathsf{H}}\mathsf{UTU}^{\mathsf{H}} = \mathsf{UT}^{\mathsf{H}}\mathsf{U}^{\mathsf{H}}\mathsf{UTU}^{\mathsf{H}} = \mathsf{UT}^{\mathsf{H}}\mathsf{TU}^{\mathsf{H}},$$

it follows from the normality of A that

$$\mathsf{TT}^{\mathsf{H}} = \mathsf{T}^{\mathsf{H}}\mathsf{T};$$

hence T is normal. From the previous lemma, we conclude that T is diagonal, and therefore, A is unitarily diagonalizable, completing the proof. □

It is rather straightforward to prove that if matrix A is normal, its singular values are the moduli of its eigenvalues.

12.5 ▪ Miscellanea on eigenvalue computations

We end this chapter by presenting a collection of results that make it possible to compute the eigenvalues, and in some cases corresponding eigenvectors, for some classes of matrices.

In general, if we know the eigenvalues of a matrix, the calculation of the corresponding eigenvectors can be done in a straightforward, albeit tedious, manner via row reduction. The calculation of the eigenvalues of a general matrix A, on the other hand, is more problematic.

Most algorithms for eigenvalue calculations proceed in an iterative manner and tend to be computationally quite demanding. In some cases, however, the calculation of the eigenvalues of a matrix is fairly straightforward. The theorems below summarize some of these special cases.

Theorem 12.23 (Eigenvalues and transposition). *The scalar $\lambda \in \mathbb{C}$ is an eigenvalue of A if and only if it is an eigenvalue of A^{T}.*

Proof. The proof of the statement follows from the fact that

$$\det(\lambda \mathsf{I}_n - \mathsf{A}^{\mathsf{T}}) = \det\left((\lambda \mathsf{I}_n - \mathsf{A})^{\mathsf{T}}\right) = \det\left(\lambda \mathsf{I}_n - \mathsf{A}\right). \qquad \square$$

The following result has already been implicitly used in the discussion, but for completeness, we state it as a theorem here.

Theorem 12.24 (Eigenvalues and eigenvectors of diagonal matrices). *The eigenvalues of a diagonal matrix $\mathsf{D} \in \mathbb{C}^{n\times n}$ are the entries d_j on its diagonal. The corresponding eigenvectors are the vectors e_j in the canonical basis of $\mathbb{R}^n$.*

Proof. The determinant of a diagonal matrix is the product of its diagonal entries. Since

$$\lambda \mathsf{I} - \mathsf{D} = \operatorname{diag}\left(\lambda - d_1, \lambda - d_2, \ldots, \lambda - d_n\right),$$

we have that

$$\det(\lambda \mathsf{I}_n - \mathsf{D}) = \prod_{j=1}^{n} (\lambda - d_j).$$

It is straightforward to check that the vector e_j is an eigenvector associated with the eigenvalue d_j. □

Theorem 12.25 (Eigenvalues of triangular matrices). *The eigenvalues of an upper triangular or lower triangular matrix* T *coincide with its diagonal entries.*

Proof. This result follows from the observation that the determinant of a triangular matrix T is the product of its diagonal entries

$$\det(\mathsf{T}) = \prod_{j=1}^{n} t_{jj}.$$

If T is triangular, then $\lambda \mathsf{I}_n - \mathsf{T}$ is also triangular with diagonal entries $\lambda - t_{jj}$; hence

$$\det(\lambda \mathsf{I}_n - \mathsf{T}) = \prod_{j=1}^{n} (\lambda - t_{jj}). \qquad \square$$

Theorem 12.26 (Eigenvalues of shifted matrices). *If* $\lambda_1, \ldots, \lambda_n$ *are the eigenvalues of* A, *then the eigenvalues of* $\mathsf{A} + \alpha \mathsf{I}_n$ *are*

$$\lambda_1 + \alpha, \ldots, \lambda_n + \alpha.$$

Proof. If v_j is an eigenvector of A corresponding to the eigenvalue λ_j, it follows from $\mathsf{A}v_j = \lambda_j v_j$ that

$$\mathsf{A}v_j + \alpha v_j = \lambda_j v_j + \alpha v_j \quad \Leftrightarrow \quad (\mathsf{A} + \alpha \mathsf{I}_n) v_j = (\lambda_j + \alpha) v_j,$$

proving that $\lambda_j + \alpha$ is an eigenvalue of $\mathsf{A} + \alpha \mathsf{I}_n$ and v_j is a corresponding eigenvector. $\square$

Theorem 12.27 (Skew symmetric matrices). *The eigenvalues of a real skew symmetric matrix are of the form* ib *with* $b \in \mathbb{R}$; *thus they are either zero or purely imaginary.*

Proof. If λ is an eigenvalue of A, then there is a vector $v \neq 0$ such that $\mathsf{A}v = \lambda v$, and

$$\langle \mathsf{A}v, v \rangle = \lambda \|v\|_2^2.$$

On the other hand, from Lemma 12.14,

$$\langle \mathsf{A}v, v \rangle = \langle v, \mathsf{A}^{\mathsf{H}} v \rangle = \langle v, -\mathsf{A}v \rangle = -\overline{\lambda} \|v\|_2^2.$$

Therefore, since $v \neq 0$,

$$\lambda = -\overline{\lambda},$$

thus the real part of λ must vanish. $\square$

Theorem 12.28 (Symmetric positive definite matrices). *A symmetric matrix* $\mathsf{A} \in \mathbb{R}^{n \times n}$ *is positive definite if and only if its eigenvalues are all positive.*

Proof. Let A be symmetric positive definite. The eigenvalues of A are real because A is real and symmetric and each eigenvalue has a real eigenvector. Let $\lambda \in \mathbb{R}$ be an eigenvalue of A, and let $\mathsf{A}v = \lambda v$ for some nonzero $v \in \mathbb{R}^n$. Then the positive definiteness of A implies that

$$v^{\mathsf{T}} \mathsf{A} v = \lambda v^{\mathsf{T}} v = \lambda \|v\|^2 > 0;$$

hence $\lambda > 0$. Conversely, if the eigenvalues of the symmetric real matrix are all positive, the eigenvalue decomposition, written as

$$\mathsf{A} = \sum_{j=1}^{n} \lambda_j v_j v_j^{\mathsf{T}},$$

implies that, for every $v \in \mathbb{R}^n$,

$$v^\mathsf{T} \mathsf{A} v = \sum_{j=1}^{n} \lambda_j (v_j^\mathsf{T} v)^2 > 0,$$

so A is positive definite. □

An immediate consequence of this result is that a symmetric positive definite matrix is invertible, its inverse is positive definite, and all its principal minors are positive definite. It can be verified that a bilinear form

$$F(v, w) = v^\mathsf{T} \mathsf{A} w,$$

where A is symmetric positive definite, defines an inner product.

Problems

1. Let A be a matrix such that the entries in each row add up to one. Show that the vector with all entries equal to 1 is an eigenvector and find the corresponding eigenvalue.

2. Show that if the entries of each row of A add up to 0, then 0 is an eigenvalue. Thus the matrix is singular.

3. Verify that if A is an $m \times n$ matrix, then $\mathsf{A}^\mathsf{T}\mathsf{A}$ is a symmetric matrix. Then show that the eigenvalues of $\mathsf{A}^\mathsf{T}\mathsf{A}$ are all nonnegative real numbers.

4. From the definition of eigenvalues and eigenvectors, and the fact that if q is an eigenvector of A associated with the eigenvalue λ, then q is in the null space of $\mathsf{A} - \lambda\mathsf{I}$, find all eigenvalues and corresponding eigenvectors of the following matrices:

$$\mathsf{A} = \begin{bmatrix} 6 & 1 & 0 & 0 \\ 0 & 3 & 1 & 0 \\ 0 & 0 & 0 & 0 \\ 0 & 0 & 0 & 2 \end{bmatrix}, \quad \mathsf{B} = \begin{bmatrix} 0 & 0 & 0 & 0 \\ 0 & 0 & 0 & 2 \\ 0 & 0 & 0 & 0 \\ 0 & 0 & 0 & -5 \end{bmatrix}, \quad \mathsf{C} = \begin{bmatrix} 6 & 1 & 0 \\ 0 & 2 & 1 \\ 0 & 0 & -2 \end{bmatrix}.$$

5. Let λ be an eigenvalue of A. Using the fact that $\det(\mathsf{A}) = \det\left(\mathsf{A}^\mathsf{T}\right)$,

   (a) show that if 0 is an eigenvalue of A, the rows of A are not linearly independent;

   (b) if the entries in each column add up to 1, show that 1 is an eigenvalue of A. Is the vector with all entries equal to 1 an eigenvector of A? Justify your answer.

6. Show that the nonzero singular values of an $m \times n$ matrix A are the square roots of the eigenvalues of $\mathsf{A}^\mathsf{T}\mathsf{A}$ and $\mathsf{A}\mathsf{A}^\mathsf{T}$.

7. Given the matrix

$$\begin{bmatrix} 1 & a & b \\ 0 & 1 & c \\ 0 & 0 & 2 \end{bmatrix},$$

   where a, b, c are arbitrary constants, what are its eigenvalues? For each eigenvalue of A, explain how its geometric multiplicity depends on the values of the constants.

8. Find by hand all eigenvalues and corresponding eigenvectors of the following matrices:

$$\mathsf{A} = \begin{bmatrix} 0 & 0 & 0 & 0 \\ 0 & 1 & 1 & 0 \\ 0 & 0 & 0 & 0 \\ 0 & 0 & 0 & 1 \end{bmatrix}, \quad \mathsf{B} = \begin{bmatrix} 0 & 0 & 0 & 0 \\ 0 & 1 & 0 & 1 \\ 0 & 0 & 0 & 0 \\ 0 & 0 & 0 & 1 \end{bmatrix}, \quad \mathsf{C} = \begin{bmatrix} 1 & 1 & 0 \\ 0 & 2 & 2 \\ 0 & 0 & 3 \end{bmatrix}.$$

9. Show that a 3×3 real matrix has at least one real eigenvalue.

10. If λ is an eigenvalue of A, it is also an eigenvalue of A^T. What can you say about the multiplicity of λ as an eigenvalue of A and as an eigenvalue of A^T? Justify your answer.

11. Given the matrix

$$\mathsf{A} = \begin{bmatrix} 2 & 1 & 0 & 0 & 0 \\ 0 & 2 & 0 & 0 & 0 \\ 0 & 0 & 1 & 1 & 0 \\ 0 & 0 & 0 & 1 & 0 \\ 0 & 0 & 0 & 0 & 3 \end{bmatrix},$$

(a) Is the matrix invertible? Justify your answer.

(b) What are its eigenvalues? List each one with its (algebraic) multiplicity.

(c) How many linearly independent eigenvectors are associated with eigenvalue 2? (It will be none if 2 is not an eigenvalue!). Find them.

12. Show that if Q is a real orthogonal 3×3 matrix, then one of its eigenvalues is either 1 or -1. So, when you rotate in three-dimensional space, one line remains unchanged, as opposed to what happens in two-dimensional space.

Chapter 13

Solution in the Least Squares Sense

In many applications, it is necessary to solve linear systems $\mathsf{A}x = b$, where A is a nonsquare matrix. In general, if the number of equations exceeds the number of degrees of freedom, the solution may not exist, and, if it exists, it may not be unique.

Therefore, asking to solve a nonsquare linear system requires that we introduce a different definition of solution. A way of generalizing the concept of solution is to look for a vector x that in some sense satisfies the equations as well as possible. This leads naturally to the concept of least squares solution. Before giving a formal definition of least squares solutions and outlining how to compute them, we introduce some special matrices, known as projectors.

13.1 ▪ Projectors

Definition 13.1. *Let* P *be an* $n \times n$ *real matrix, and let* $V \subset \mathbb{R}^n$ *be the subspace spanned by its columns. The matrix* P *is a* projector *onto* V *if it satisfies the condition*

$$\mathsf{PP} = \mathsf{P}.$$

A projector P *is an* orthogonal projector *if it is symmetric,*

$$\mathsf{P}^\mathsf{T} = \mathsf{P}.$$

Theorem 13.2. *The eigenvalues of a projector* $\mathsf{P} \in \mathbb{R}^{n\times n}$ *are equal to* 1 *or* 0.

Proof. Since P is a projector, we have

$$\mathsf{P}^2 = \mathsf{P}.$$

If λ is an eigenvalue of P and v is an associated eigenvector, then

$$\lambda v = \mathsf{P}v = \mathsf{P}^2 v = \mathsf{P}(\mathsf{P}v) = \lambda \mathsf{P}v = \lambda^2 v.$$

Since $v \neq 0$,

$$\lambda v = \lambda^2 v \Rightarrow \lambda = \lambda^2 \Rightarrow \lambda(\lambda - 1) = 0;$$

therefore either $\lambda = 0$ or $\lambda = 1$. $\square$

Example 13: Let $V, W \subset \mathbb{R}^n$ be two subspaces such that

$$\mathbb{R}^n = V \oplus W.$$

The projector P on V can be defined as follows: For $x \in \mathbb{R}^n$,

$$x = v + w, \qquad \mathsf{P}x = v;$$

that is, P picks the unique component of x that lies in the subspace V. In general, a projector constructed in this way need not be orthogonal, unless the two subspaces V and W are mutually orthogonal.

Observe that if $\mathsf{P} \in \mathbb{R}^{n\times n}$ is a projector, the matrix $\mathsf{I} - \mathsf{P}$ is also a projector since

$$(\mathsf{I} - \mathsf{P})(\mathsf{I} - \mathsf{P}) = \mathsf{I} - \mathsf{P} - \mathsf{P} + \mathsf{PP} = \mathsf{I} - \mathsf{P}.$$

Furthermore, if P is an orthogonal projector, the ranges of P and $\mathsf{I} - \mathsf{P}$ are mutually orthogonal. In fact, if $x \in \mathbb{R}^n$ is in the range of P, and $z \in \mathbb{R}^n$ is in the range of $\mathsf{I} - \mathsf{P}$, we have

$$x = \mathsf{P}u, \qquad z = (\mathsf{I} - \mathsf{P})v$$

for some $u, v \in \mathbb{R}^n$. Then,

$$\begin{aligned} x^{\mathsf{T}} z &= (\mathsf{P}u)^{\mathsf{T}}(\mathsf{I} - \mathsf{P})v = u^{\mathsf{T}}(\mathsf{P}^{\mathsf{T}} - \mathsf{P}^{\mathsf{T}}\mathsf{P})v \\ &= u^{\mathsf{T}}(\mathsf{P} - \mathsf{PP})v = u^{\mathsf{T}}(\mathsf{P} - \mathsf{P})v = 0 \end{aligned}$$

for all u, v, proving the orthogonality. Notice that each $x \in \mathbb{R}^n$ can be written as

$$x = \mathsf{P}x + (\mathsf{I} - \mathsf{P})x,$$

giving rise to the orthogonal decomposition

$$\mathbb{R}^n = \mathcal{R}(\mathsf{P}) \oplus \mathcal{R}(\mathsf{I} - \mathsf{P}).$$

The definition of projector assumes that the matrix P is given. Usually, however, we start with a subspace and need to construct the projector onto the subspace. The following theorem gives an explicit expression for an orthogonal projector onto a given subspace.

Theorem 13.3. *A projector P onto a subspace $V \subset \mathbb{R}^n$ of dimension k is a matrix of rank k.*

1. *If $q_1, q_2, \ldots, q_k$ is an orthonormal basis of V, and*

$$\mathsf{Q} = \begin{bmatrix} \vdots & & \vdots \\ q_1 & \cdots & q_k \\ \vdots & & \vdots \end{bmatrix} \in \mathbb{R}^{n\times k},$$

the orthogonal projector onto V is given by

$$\mathsf{P} = \mathsf{QQ}^{\mathsf{T}}.$$

2. *More generally, if $a_1, a_2, \ldots, a_k$ is a basis of V and*

$$\mathsf{A} = \begin{bmatrix} \vdots & & \vdots \\ a_1 & \cdots & a_k \\ \vdots & & \vdots \end{bmatrix} \in \mathbb{R}^{n\times k},$$

then the orthogonal projector on V is given by

$$\mathsf{P} = \mathsf{A}\left(\mathsf{A}^{\mathsf{T}}\mathsf{A}\right)^{-1}\mathsf{A}^{\mathsf{T}}.$$

Proof. First, we observe that if P is a projector onto V, we have $\mathcal{R}(\mathsf{P}) = V$, and therefore the columns of P span a k-dimensional subspace, implying that P is a rank-k matrix.

1. Since the columns of Q can be completed into an orthonormal basis $\{q_1, q_2, \dots, q_n\}$ of $\mathbb{R}^n$, every vector $x \in \mathbb{R}^n$ can be written as

$$x = \sum_{j=1}^{n} \gamma_j q_j = \sum_{j=1}^{k} \gamma_j q_j + \sum_{j=k+1}^{n} \gamma_j q_j = v + w,$$

where

$$v \in \mathcal{R}(\mathsf{Q}), \quad w \perp \mathcal{R}(\mathsf{Q}).$$

By the orthonormality of the basis, we have

$$\gamma_j = q_j^\mathsf{T} x, \quad 1 \le j \le n;$$

therefore the orthogonal projection $\mathsf{P}x$ of x onto the column space of Q can be written as

$$\mathsf{P}x = \sum_{j=1}^{k} (q_j^\mathsf{T} x) q_j = \left(\sum_{j=1}^{k} q_j q_j^\mathsf{T} \right) x = \mathsf{Q}\mathsf{Q}^\mathsf{T} x.$$

From the orthonormality of the vectors q_k it follows that

$$(\mathsf{Q}\mathsf{Q}^\mathsf{T})(\mathsf{Q}\mathsf{Q}^\mathsf{T}) = \mathsf{Q}(\mathsf{Q}^\mathsf{T}\mathsf{Q})\mathsf{Q}^\mathsf{T} = \mathsf{Q}\mathsf{I}_k\mathsf{Q}^\mathsf{T} = \mathsf{Q}\mathsf{Q}^\mathsf{T},$$

thus proving that $\mathsf{Q}\mathsf{Q}^\mathsf{T}$ is a projector. Moreover, since $\mathsf{Q}\mathsf{Q}^\mathsf{T}$ is symmetric, it is an orthogonal projector.

2. We use the QR-factorization to prove the second part: Let

$$\mathsf{A} = \mathsf{Q}\mathsf{R},$$

where $\mathsf{Q} \in \mathbb{R}^{n \times n}$ is orthogonal, $\mathsf{R} \in \mathbb{R}^{n \times k}$ is upper triangular, and

$$\mathrm{span}\,\{a_1, a_2, \dots, a_k\} = \mathrm{span}\,\{q_1, q_2, \dots, q_k\}.$$

If we denote by $\mathsf{Q}_k \in \mathbb{R}^{n \times k}$ the matrix with columns $q_1, \dots, q_k$, it follows from part 1 that

$$\mathsf{P} = \mathsf{Q}_k \mathsf{Q}_k^\mathsf{T}.$$

On the other hand, if we partition R as

$$\mathsf{R} = \left[\frac{\mathsf{R}_k}{\mathsf{O}_{(n-k)\times k}} \right],$$

where R_k is the upper triangular $k \times k$ principal minor of R,

$$\mathsf{A} = \mathsf{Q}_k \mathsf{R}_k.$$

Since the columns of A are linearly independent, the matrix R_k is invertible and

$$\mathsf{Q}_k = \mathsf{A}\mathsf{R}_k^{-1}.$$

Moreover,

$$\mathsf{A}^{\mathsf{T}}\mathsf{A} = \mathsf{R}_k^{\mathsf{T}}\mathsf{Q}_k^{\mathsf{T}}\mathsf{Q}_k\mathsf{R}_k = \mathsf{R}_k^{\mathsf{T}}\mathsf{R}_k$$

is invertible, and

$$\left(\mathsf{A}^{\mathsf{T}}\mathsf{A}\right)^{-1} = \mathsf{R}_k^{-1}\mathsf{R}_k^{-\mathsf{T}}.$$

Therefore

$$\begin{aligned}\mathsf{P} &= \mathsf{Q}_k\mathsf{Q}_k^{\mathsf{T}}\\ &= \mathsf{A}\mathsf{R}_k^{-1}\mathsf{R}_k^{-\mathsf{T}}\mathsf{A}^{\mathsf{T}}\\ &= \mathsf{A}\left(\mathsf{R}_k^{\mathsf{T}}\mathsf{R}_k\right)^{-1}\mathsf{A}^{\mathsf{T}}\\ &= \mathsf{A}\left(\mathsf{A}^{\mathsf{T}}\mathsf{A}\right)^{-1}\mathsf{A}^{\mathsf{T}},\end{aligned}$$

completing the proof of part 2. □

Above, the projector P was constructed by using the first k of the orthonormal vectors q_j. In general, given an orthonormal basis $\{q_1, q_2, \dots, q_n\}$ of $\mathbb{R}^n$, let $\mathsf{Q} \in \mathbb{R}^{n\times n}$ be the orthogonal matrix with columns q_j, and for any k, $k \le n$, we may partition the matrix $\mathsf{Q} \in \mathbb{R}^{n\times n}$ column-wise as

$$\mathsf{Q} = \left[\begin{array}{cc}\mathsf{Q}_k & \mathsf{Q}_k'\end{array}\right], \quad \mathsf{Q}_k \in \mathbb{R}^{n\times k}, \quad \mathsf{Q}_k' \in \mathbb{R}^{n\times(n-k)},$$

and define two orthogonal projectors,

$$\mathsf{P}_1 = \mathsf{Q}_k\mathsf{Q}_k^{\mathsf{T}}, \quad \mathsf{P}_2 = \mathsf{Q}_k'(\mathsf{Q}_k')^{\mathsf{T}}.$$

Then,

$$\mathsf{P}_1 + \mathsf{P}_2 = \mathsf{I},$$

as can be easily verified. This way of constructing projectors on orthogonal subspaces is used in the following discussion.

13.2 ▪ Least squares solution

Consider a linear system $\mathsf{A}x = b$ with coefficient matrix $\mathsf{A} \in \mathbb{R}^{m\times n}$, $m > n$, and the right-hand side $b \in \mathbb{R}^m$. If $b \notin \mathcal{R}(\mathsf{A})$, there is no vector x such that $\mathsf{A}x = b$, because in that case b would be in the range of A. Therefore, the concept of a solution needs to be generalized. We start with the following definition.

Definition 13.4. *Given a matrix* $\mathsf{A} \in \mathbb{R}^{m\times n}$ *and a vector* $b \in \mathbb{R}^m$, *a vector* x_{LS} *is a* least squares solution *of the linear system* $\mathsf{A}x = b$ *if it minimizes the expression*

$$F(x) = \|b - \mathsf{A}x\|^2.$$

The name least squares solution refers to the fact that it minimizes the sum of squares of the components of the residual vector,

$$r = b - \mathsf{A}x. \tag{13.1}$$

If $b \in \mathcal{R}(\mathsf{A})$, there exists a vector x such that $b = \mathsf{A}x$ and the residual error is the zero vector, which is the minimum possible value of the function to minimize.

The least squares solution need not be unique. In fact, it is easy to construct examples for which there are infinitely many minimizers of $F(x)$.

The following theorem gives us a characterization of the least squares solution that can be used for its computation, as well as the conditions guaranteeing its existence and uniqueness.

Theorem 13.5. *The vector x solves the linear system $\mathsf{A}x = b$ in the least squares sense if and only if*

$$(b - \mathsf{A}x) \perp \mathcal{R}(\mathsf{A}). \tag{13.2}$$

If the columns of the matrix A are linearly independent, the least squares solution is unique, and it is given by

$$x_{\mathrm{LS}} = (\mathsf{A}^{\mathsf{T}}\mathsf{A})^{-1}\mathsf{A}^{\mathsf{T}}b. \tag{13.3}$$

Proof. Consider the orthogonal decomposition of the image space of A,

$$\mathbb{R}^m = \mathcal{R}(\mathsf{A}) \oplus \mathcal{N}(\mathsf{A}^{\mathsf{T}}),$$

and write the corresponding orthogonal decomposition of b,

$$b = \mathsf{P}_1 b + \mathsf{P}_2 b,$$

where P_1 is the orthogonal projector onto $\mathcal{R}(\mathsf{A})$ and $\mathsf{P}_2 = \mathsf{I} - \mathsf{P}_1$ is the orthogonal projection onto $\mathcal{N}(\mathsf{A}^{\mathsf{T}})$. This induces an orthogonal decomposition of the residual (13.1),

$$r = b - \mathsf{A}x = (\mathsf{P}_1 b - \mathsf{A}x) + \mathsf{P}_2 b = r_1 + r_2,$$

where $r_1 \in \mathcal{R}(\mathsf{A})$, $r_2 \in \mathcal{N}(\mathsf{A}^{\mathsf{T}})$. The orthogonality of r_1 and r_2 implies that

$$\begin{aligned} F(x) = \|r\|^2 &= (r_1 + r_2)^{\mathsf{T}} (r_1 + r_2) \\ &= r_1^{\mathsf{T}} r_1 + r_1^{\mathsf{T}} r_2 + r_2^{\mathsf{T}} r_1 + r_2^{\mathsf{T}} r_2 \\ &= \|r_1\|^2 + \|r_2\|^2, \end{aligned}$$

and

$$F(x) \geq \|r_2\|^2,$$

with equality holding if x satisfies

$$r_1 = \mathsf{P}_1 b - \mathsf{A}x = 0. \tag{13.4}$$

Since $\mathsf{P}_1 b \in \mathcal{R}(\mathsf{A})$, there is at least one solution of (13.4). Therefore, for any least squares solution vector, the residual vector satisfies

$$r = b - \mathsf{A}x = r_2 \perp \mathcal{R}(\mathsf{A}),$$

thus proving the first part of the claim.

To prove the second part, assume that the columns of A are linearly independent. It follows from Theorem 13.3 that

$$r_1 = \mathsf{A}(\mathsf{A}^{\mathsf{T}}\mathsf{A})^{-1}\mathsf{A}^{\mathsf{T}}b - \mathsf{A}x = \mathsf{A}\left((\mathsf{A}^{\mathsf{T}}\mathsf{A})^{-1}\mathsf{A}^{\mathsf{T}}b - x\right);$$

hence condition (13.4) is satisfied if

$$x = (\mathsf{A}^{\mathsf{T}}\mathsf{A})^{-1}\mathsf{A}^{\mathsf{T}}b,$$

proving the second claim. □

Notice that the linear independence of the columns of A leading to formula (13.3) implies that the null space of A is trivial. If the columns of A are not linearly independent, which happens in particular when $m < n$, the matrix A has a nontrivial null space. If $x_{\rm LS}$ is one least squares solution and $x_0 \in \mathcal{N}(\mathsf{A})$, $x_0 \neq 0$, then

$$F(x_{\rm LS}) = F(x_{\rm LS} + x_0);$$

therefore the least squares solution cannot be unique. To single out one solution, we give the following definition.

Definition 13.6. *Let* $\mathsf{A} \in \mathbb{R}^{m\times n}$ *and* $b \in \mathbb{R}^n$. *The least squares solution* $x_{\rm MN}$ *to the problem* $\mathsf{A}x = b$ *that satisfies*

$$\|x_{\rm MN}\| \leq \|x_{\rm LS}\| \quad \textit{for all least squares solutions } x_{\rm LS}$$

is called the minimum norm solution. *The minimum norm solution is characterized by the property*

$$x_{\rm MN} \perp \mathcal{N}(\mathsf{A}).$$

A least squares solution does not guarantee in general an exact match of the right-hand side vector b with $\mathsf{A}x_{\rm LS}$; however, the error that remains is in the subspace orthogonal to $\mathcal{R}(\mathsf{A})$. The residual norm,

$$\|b - \mathsf{A}x_{\rm LS}\| = \|(\mathsf{I} - \mathsf{P})b\|,$$

where $\mathsf{P} : \mathbb{R}^m \to \mathcal{R}(\mathsf{A})$ is the orthogonal projection, measures the distance between b and the range of A. If $b \in \mathcal{R}(\mathsf{A})$, then $(\mathsf{I} - \mathsf{P})b = 0$, and $b = \mathsf{A}x_{\rm LS}$; that is, if the linear system has a solution, that solution is the least squares solution also.

An alternative way to arrive at the least squares solution (13.3) is to notice that condition (13.2) is equivalent to requiring the residual to be orthogonal to each column a_j of A, that is,

$$a_j^{\mathsf{T}}(b - \mathsf{A}x) = 0, \quad 1 \leq j \leq n.$$

Expressing the aggregate of these conditions in matrix form yields the linear system

$$\mathsf{A}^{\mathsf{T}}(b - \mathsf{A}x) = 0,$$

or, equivalently,

$$\mathsf{A}^{\mathsf{T}}\mathsf{A}x = \mathsf{A}^{\mathsf{T}}b. \tag{13.5}$$

The latter is referred to as the *normal equations* of the least squares problem.

Finally, if A has linearly independent columns, the matrix mapping the data b to the least squares solution,

$$\mathsf{A}^{\dagger} = \left(\mathsf{A}^{\mathsf{T}}\mathsf{A}\right)^{-1}\mathsf{A}^{\mathsf{T}} \in \mathbb{R}^{n\times m},$$

is called the *pseudoinverse* of A. Observe that if $m = n$ and A is invertible, the pseudoinverse coincides with the inverse.

13.3 ▪ Computing the least squares solution

There are different strategies to compute the least squares solution of a linear system, some of which may be preferable to the others. Here we briefly outline three strategies, without discussing their relative pros and cons.

13.4 ▪ Least squares via SVD

Assume that the singular value decomposition of A is available. If r is the rank of the matrix A, hence the number of nonzero singular values, we may write A in terms of its thin SVD as

$$\mathsf{A} = \mathsf{U}\Sigma\mathsf{V}^\mathsf{T} = \mathsf{U}_r\Sigma_r\mathsf{V}_r^\mathsf{T},$$

where $\mathsf{U}_r \in \mathbb{R}^{m\times r}$, $\mathsf{V}_r \in \mathbb{R}^{n\times r}$, and $\Sigma_r \in \mathbb{R}^{r\times r}$. Let us partition the matrix U as

$$\mathsf{U} = \left[\begin{array}{cc} \mathsf{U}_r & \mathsf{U}_r' \end{array}\right], \quad \mathsf{U}_r \in \mathbb{R}^{m\times r}, \quad \mathsf{U}_r' \in \mathbb{R}^{m\times(m-r)},$$

and write the linear system $\mathsf{A}x - b$ as

$$\begin{aligned} \mathsf{U}_r\Sigma_r\mathsf{V}_r^\mathsf{T}x &= \mathsf{U}_r\mathsf{U}_r^\mathsf{T}b + \mathsf{U}_r'(\mathsf{U}_r')^\mathsf{T}b \\ &= b_1 + b_2, \end{aligned}$$

where $b_1 \in \mathcal{R}(\mathsf{A})$ and $b_2 \perp \mathcal{R}(\mathsf{A})$.

Solving the linear system in the least squares sense is tantamount to solving exactly the linear system for the component of the right-hand side in the column space of A, that is,

$$\mathsf{U}_r\Sigma_r\mathsf{V}_r^\mathsf{T}x = \mathsf{U}_r\mathsf{U}_r^\mathsf{T}b,$$

because b_2 cannot be approximated in any way from within the range of A.

Multiplying both sides of the above equation by U_r^T and using the orthogonality of the columns of U_r, we obtain

$$\underbrace{(\mathsf{U}_r^\mathsf{T}\mathsf{U}_r)}_{=\mathsf{I}_r}\Sigma_r\mathsf{V}_r^\mathsf{T}x = \underbrace{(\mathsf{U}_r^\mathsf{T}\mathsf{U}_r)}_{=\mathsf{I}_r}\mathsf{U}_r^\mathsf{T}b,$$

and letting $y = \mathsf{V}_r^\mathsf{T}x \in \mathbb{R}^r$, $z = \mathsf{U}_r^\mathsf{T}b \in \mathbb{R}^r$, we arrive at the system

$$\Sigma_r y = z \Leftrightarrow y = \Sigma_r^{-1}z.$$

If the columns of A are linearly independent, the matrix A has full column rank; thus $r = n$ and $\mathsf{V}_r = \mathsf{V}$, and we find the solution

$$x = \mathsf{V}y = \mathsf{V}\Sigma_n^{-1}\mathsf{U}_n^\mathsf{T}b.$$

The SVD analysis of a nonsquare linear system sheds light on what happens when $r < n$: If $x \in \mathbb{R}^n$ is expressed as a linear combination of the columns of V, partitioned as

$$\mathsf{V} = \left[\begin{array}{cc} \mathsf{V}_r & \mathsf{V}_r' \end{array}\right], \quad \mathsf{V}_r \in \mathbb{R}^{n\times r}, \quad \mathsf{V}_r' \in \mathbb{R}^{n\times(n-r)},$$

writing

$$x = \mathsf{V}_r\alpha + \mathsf{V}_r'\beta, \quad \alpha \in \mathbb{R}^r, \quad \beta \in \mathbb{R}^{n-r},$$

the condition $y = \mathsf{V}_r^\mathsf{T}x$ determines only the entries of the vector α, while the entries of β can be chosen arbitrarily. Indeed,

$$y = \mathsf{V}_r^\mathsf{T}x = \mathsf{V}_r^\mathsf{T}(\mathsf{V}_r\alpha + \mathsf{V}_r'\beta) = \alpha,$$

and

$$x = \mathsf{V}_r\Sigma_r^{-1}\mathsf{U}_r^\mathsf{T}b + \mathsf{V}_r'\beta, \quad \beta \in \mathbb{R}^{n-r} \text{ arbitrary}.$$

In this case, the least squares solution is nonunique: the minimum norm solution is obtained by setting $\beta = 0$.

We summarize these results in the following theorem.

Theorem 13.7. *Let* $\mathsf{A} \in \mathbb{R}^{m\times n}$ *be a matrix of rank* r*, and let*

$$\mathsf{A} = \mathsf{U}\Sigma\mathsf{V}^{\mathsf{T}}$$

be its singular value decomposition.

The general least squares solution of the problem $\mathsf{A}x = b$ *is*

$$x_{\mathrm{LS}} = \mathsf{V}_r\Sigma_r^{-1}\mathsf{U}_r^{\mathsf{T}}b + \mathsf{V}_r'\beta$$

for any $\beta \in \mathbb{R}^{n-r}$.

If the columns of A *are linearly independent,* $r = n$*, and the least squares solution is unique,*

$$x_{\mathrm{LS}} = \mathsf{V}\Sigma_n^{-1}\mathsf{U}_n^{\mathsf{T}}b.$$

If $r < n$*, the least squares solution is nonunique, and the solution* x_{MN} *of minimum norm is*

$$x_{\mathrm{MN}} = \mathsf{V}_r\Sigma_r^{-1}\mathsf{U}_r^{\mathsf{T}}b.$$

13.5 ▪ Least squares via QR

The QR-factorization of the matrix A provides another way to solve the least squares problem. If $m \geq n$,

$$\mathsf{A} = \mathsf{QR} = \begin{bmatrix} \mathsf{Q}_n & \mathsf{Q}_n' \end{bmatrix} \begin{bmatrix} \widetilde{\mathsf{R}} \\ \mathsf{O}_{(m-n)\times n} \end{bmatrix},$$

where $\widetilde{\mathsf{R}} \in \mathbb{R}^{n\times n}$ is an upper triangular matrix. If the columns of A are linearly independent, $\widetilde{\mathsf{R}}$ is invertible, and the least squares problem can be written as

$$\begin{bmatrix} \mathsf{Q}_n & \mathsf{Q}_n' \end{bmatrix} \begin{bmatrix} \widetilde{\mathsf{R}} \\ \mathsf{O}_{(m-n)\times n} \end{bmatrix} x = \begin{bmatrix} b_1 \\ b_2 \end{bmatrix},$$

where $b_1 \in \mathbb{R}^n$, $b_2 \in \mathbb{R}^{(m-n)}$. Since multiplication by orthogonal matrices preserves the Euclidian norm, we can write

$$\begin{aligned} \|\mathsf{A}x - b\| &= \|\mathsf{Q}^{\mathsf{T}}(\mathsf{A}x - b)\| \\ &= [\mathsf{Q}^{\mathsf{T}}\mathsf{QR}x - \mathsf{Q}^{\mathsf{T}}b\| \\ &= \left\| \begin{bmatrix} \widetilde{\mathsf{R}} \\ \mathsf{O}_{(m-n)\times n} \end{bmatrix} x - \mathsf{Q}^{\mathsf{T}}b \right\| \\ &= \left\| \begin{bmatrix} \widetilde{\mathsf{R}}x \\ 0 \end{bmatrix} - \begin{bmatrix} \mathsf{Q}_n^{\mathsf{T}}b \\ (\mathsf{Q}_n')^{\mathsf{T}}b \end{bmatrix} \right\|. \end{aligned}$$

Since the only way to reduce the norm of the residual is to solve

$$\widetilde{\mathsf{R}}x = \mathsf{Q}_n^{\mathsf{T}}b,$$

the least squares solution is

$$x_{\mathrm{LS}} = \widetilde{\mathsf{R}}^{-1}\mathsf{Q}_n^{\mathsf{T}}b.$$

13.6 ▪ Least squares via Gaussian elimination

Because the least squares solution of the linear system is the solution of the normal equations,

$$\mathsf{A}^\mathsf{T}\mathsf{A}x = \mathsf{A}^\mathsf{T}b,$$

we can compute the LU factorization of $\mathsf{A}^\mathsf{T}\mathsf{A}$ and solve two triangular linear systems.

If the columns of A are linearly independent, the matrix $\mathsf{A}^\mathsf{T}\mathsf{A}$ is invertible; hence the solution of the normal equations exists and is unique. Moreover, in that case the matrix $\mathsf{A}^\mathsf{T}\mathsf{A}$ is symmetric positive definite and hence admits a Cholesky factorization.

In particular, if $\mathsf{A} = \mathsf{QR}$ is the QR factorization of A, then

$$\mathsf{A}^\mathsf{T}\mathsf{A} = \mathsf{R}^\mathsf{T}\mathsf{R};$$

thus R is the Cholesky factor of $\mathsf{A}^\mathsf{T}\mathsf{A}$.

Mathematically the three ways to compute the solution of the linear system in the least squares sense are equivalent; however numerical consideration may favor one of the three methods over the others.

Problems

1. Recalling the definition of least squares solution, if

$$\mathsf{A} = \begin{bmatrix} 1 & -1 & 1 \\ 1 & 0 & 0 \\ 1 & 1 & 1 \\ 1 & 2 & 2 \end{bmatrix}, \qquad b = \begin{bmatrix} 8 \\ 8 \\ 4 \\ 16 \end{bmatrix},$$

   decide, without solving the problem, which of the following three vectors could be the least squares solution to $\mathsf{A}x = b$:

$$\begin{bmatrix} 5 \\ -1 \\ 4 \end{bmatrix}, \quad \begin{bmatrix} 8 \\ 1 \\ -2 \end{bmatrix}, \quad \begin{bmatrix} 0 \\ -1 \\ 0 \end{bmatrix}.$$

2. Consider a subspace V of $\mathbb{R}^n$ of dimension m. If the $n \times n$ matrix A is the orthogonal projector onto V, what can you say about the eigenvalues and corresponding eigenvectors of A? What is the algebraic and geometric multiplicity of each eigenvalue? You can use a geometric argument.

3. If

$$\mathsf{A} = \begin{bmatrix} 1 & 1 \\ 1 & 0 \\ 1 & 1 \end{bmatrix}, \qquad b = \begin{bmatrix} 3 \\ 3 \\ 3 \end{bmatrix},$$

   find the projection of b onto the range of A and the projector onto the range of A.

4. Find the least squares solution of $\mathsf{A}x = b$ for

$$\mathsf{A} = \begin{bmatrix} 1 \\ 2 \\ 3 \end{bmatrix}, \qquad b = \begin{bmatrix} 2 \\ 3 \\ 7 \end{bmatrix}.$$

Chapter 14

Further Reading

Linear algebra is a very rich branch of mathematics with applications to many different areas. As the goal of this book was to provide a concise introduction to linear algebra, several topics had to be omitted. There are many excellent textbooks for a first course in linear algebra where readers can find proofs and worked-out examples illustrating the concepts introduced; see, e.g., the books by Gil Strang [1] and Elizabeth and Mark Meckes [2]. For readers longing for a more theoretical or for a more computational approach, there are numerous monographs on linear algebra that can be used to deepen knowledge and explore how operations should be organized when they are to be performed on a computer.

Among the classic theoretical linear algebra books, Paul Halmos' *Finite-Dimensional Vector Spaces* [3] focuses on vector spaces, while the emphasis of *Matrix Analysis* by Roger Horn and Charles Johnson is more on matrices [4]. Another noteworthy reference on matrix analysis and linear algebra is the book of Carl Meyer [5]. The comprehensive treatise *Linear Algebra* by Kenneth Hoffman and Ray Kunze [6], where everything is proved in a careful and uncompromising way, is a classic text in linear algebra that may provide the answer to many theoretical questions that were not treated here.

With the use of digital computers for calculations, it became necessary to address how linear algebra operations should be organized to maximize accuracy and computational efficiency, making numerical linear algebra a central area of applied mathematics. While the answer, or at least a reference, for most numerical linear algebra questions can be found in the encyclopedic *Matrix Computations* [7] book by Gene Golub and Charles Van Loan, the cut and dry style of the presentation may be discouraging for people entering the field. Readers looking for a comprehensive introduction to numerical linear algebra may want to consider *Numerical Linear Algebra* by Lloyd Trefethen and David Bau [8] and *Applied Numerical Linear Algebra* by James Demmel [9]. Both books provide an extensive introduction to how to organize linear algebra computations and an explanation of how sometimes theory and computation seem to give conflicting results. For an in-depth treatment of the accuracy and stability of algorithms, we refer the reader to the comprehensive book by Nick Higham [10]. Readers who adhere to the the less is more philosophy may appreciate the approach to numerical linear algebra in the two books by Alan Laub [11, 12].

[1] G. Strang, *Introduction to Linear Algebra*, 5th ed., Wellesley-Cambridge Press, Wellesley, MA, 2016.

[2] E.S. Meckes and M.W. Meckes, *Linear Algebra*, Cambridge University Press, Cambridge, UK, 2018.

[3] P.H. Halmos, *Finite-Dimensional Vector Spaces*, Courier Dover Publications, Mineola, NY, 2017.

[4] R.A. Horn and C.R. Johnson, *Matrix Analysis*, Cambridge University Press, Cambridge, UK, 2012.

[5] C. Meyer, *Matrix Analysis and Applied Linear Algebra*, SIAM, Philadelphia, 2000.

[6] K.M. Hoffman and R. Kunze, *Linear Algebra*, 2nd ed., Prentice-Hall, Englewood Cliffs, NJ, 1971.

[7] G.H. Golub and C.F. Van Loan, *Matrix Computations*, 4th ed., Johns Hopkins University Press, Baltimore, MD, 2013.

[8] L.N. Trefethen and D. Bau, III, *Numerical Linear Algebra*, SIAM, Philadelphia, 1997.

[9] J.W. Demmel, *Applied Numerical Linear Algebra*, SIAM, Philadelphia, 1997.

[10] N.J. Higham, *Accuracy and Stability of Numerical Algorithms*, 2nd ed., SIAM, Philadelphia, 2002.

[11] A. Laub, *Computational Matrix Analysis*, SIAM, Philadelphia, 2012.

[12] A. Laub, *Matrix Analysis for Scientists and Engineers*, SIAM, Philadelphia, 2005.

Index